4th EDITION
DATELINE CANADA
UNDERSTANDING ECONOMICS THROUGH PRESS REPORTS

Peter Kennedy
Simon Fraser University

Gary Dorosh
Douglas College

PRENTICE-HALL CANADA INC., SCARBOROUGH, ONTARIO

To DATE
Who wishes she could understand,

and

To DAD
Who always knew I could do it.

Canadian Cataloguing in Publication Data

Kennedy, Peter, 1943-
  Dateline Canada : understanding economics through press reports

ISBN 0-13-200361-9

1. Economics. 2. Canada—Economic conditions—1971-* I. Dorosh, Gary, 1942- II. Title.

HB171.5.K46 1989   330   C89-094722-8

©1990 Prentice-Hall Canada Inc.
Scarborough, Ontario

All rights reserved. No part of this book may be reproduced in any form without permission in writing from the publishers.

Prentice-Hall Inc., *Englewood Cliffs, New Jersey*
Prentice-Hall International, Inc., *London*
Prentice-Hall of Australia, Pty., *Sydney*
Prentice-Hall of India Pvt., Ltd., *New Delhi*
Prentice-Hall of Japan, Inc., *Tokyo*
Prentice-Hall of Southeast Asia (Pte.) Ltd., *Singapore*
Editoria Prentice-Hall do Brasil Ltda., *Rio de Janeiro*
Prentice-Hall Hispanoamericana, S.A., *Mexico*
Whitehall Books Limited, *Wellington, New Zealand*

ISBN 0-13-200361-9

1 2 3 4 5   WC   93 92 91 90 89

Typesetting by CompuScreen Typesetting Ltd.
Printed and bound in Canada by Webcom

Cover and Interior Design: Denise Marcella
Cover Layout: Alex Baldocchi
Production Editors: Ted Emerson/Doris Wolf
Production Coordinator: Sandra Paige

Contents

Preface to the Instructor vii

Preface to the Student ix

Cross-References x

PART I   MICROECONOMICS 1

I-1   Supply and Demand 2

    A Oil 3
    B Newsprint 4
    C Pulp 5
    D Antifreeze 6
    E Beer 7
       Short Clippings 8

I-2   Elasticity 18

    A Natural Gas 19
    B Software 20
    C Gasoline 21
       Short Clippings 23

I-3   Costs and Perfect Competition 28

    A Ranching 29
    B Forests 30
       Short Clippings 31

I-4   The Corn-Hog Cycle 40

    A Newsprint 41
    B Beef 42
       Short Clippings 43

I-5   Marketing Boards 46

    A Poultry and Eggs 47
    B Milk 49
       Short Clippings 51

I-6   Government-Fixed Prices 56

    A Minimum Wages 57
    B Rent Controls: Pro 58
       Short Clippings 60

I-7 At the Margin 66
- A Cross-Subsidization 67
- B Electric Metering 68
- Short Clippings 70

I-8 Taxes and Subsidies 74
- A Pollution 76
- B Day Care 77
- C Government Cafeterias 78
- Short Clippings 80

I-9 Foreign Competition 86
- A Lumber 87
- B Apples 88
- C Tariffs vs Quotas 89
- Short Clippings 91

I-10 Imperfect Competition 100
- A Superstores 101
- B Blue Jeans 102
- Short Clippings 104

I-11 Regulated Monopolies 108
- A Railroads 109
- B Telephones 111
- Short Clippings 112

I-12 Price Discrimination 114
- A First Class Travel 115
- B Excursion Rates 116
- Short Clippings 117

I-13 Government Franchises 118
- A Freeway Gas Stations 119
- B Duty Free Shops 120
- Short Clippings 122

PART II MACROECONOMICS 125

II-1 Unemployment 126
- A Changing Unemployment 128
- B High Wages 129
- Short Clippings 130

II-2 Consumption and Aggregate Demand 138

    A Uncertain Saving   139
    B GDP Rise   140
        Short Clippings   140

II-3 Investment and the Inventory Cycle 146

    A Recovery Stronger   147
    B Investment Surge   148
        Short Clippings   150

II-4 Fiscal Policy 156

    A Using Fiscal Policy   157
        Short Clippings   159

II-5 Monetary Policy 166

    A T-Bill Auction   168
    B Struggling Monetarism   169
        Short Clippings   171

II-6 Inflation 182

    A Measuring Inflation   183
    B Inflation Fight   184
        Short Clippings   186

II-7 Interest Rates 192

    A Challenging Conventional Wisdom   193
    B Investment Planning   194
        Short Clippings   196

II-8 International Influences 206

    A Dollar-Defense Fund   209
    B Big Mac Index   210
    C Made-in-Canada Monetary Policy   211
        Short Clippings   213

II-9 Stagflation 232

    A The Natural Rate of Unemployment   234
    B Unemployment versus Inflation   235
        Short Clippings   236

II-10 Government Deficits 246

    A Debt Problem 247
    B Costs vs Benefits of Deficit 248
       Short Clippings 250

II-11 Free Trade 258

    A Free Trade: Pro 259
    B Free Trade: Con 260
       Short Clippings 261

APPENDIX Suggested Answers to Selected Questions **263**

GLOSSARY 301

Preface to the Instructor

The main purpose of this book is to bridge the gap between economic theory and the real world by asking readers to interpret and criticize the economic content of newspaper clippings. In addition to illustrating real-world applications of economic theory, these clippings provide a wealth of institutional information about the Canadian economy which we personally find to be of considerable value. Although the book has been designed for the economics-principles student, its contents range quite widely in topic coverage and degree of difficulty, so there is plenty of scope for its use in courses that follow the principles course.

Several significant changes have been made in this fourth edition. As would be expected, outdated articles have been replaced with new ones, and currently popular topics, such as government budget deficits and free trade, have been strengthened. The biggest change, however, is a restructuring of the short clippings at the end of each chapter; they are now arranged in subsections, by topic, with about seven or eight clippings in each subsection. This should make it easier to identify clippings related to specific points of interest.

In our view, the short clippings are the strongest feature of this book. (There are now more than 650 short clippings, about 60% in the macroeconomics half.) They provide far more variety than full-length articles, and allow you to tailor your use of this book to match more closely your interests and your students' abilities. We are indebted to several Canadian newspapers, in particular the *Financial Post*, the *Financial Times of Canada*, the *Toronto Globe and Mail*, the *Toronto Star*, the *Vancouver Sun* and the *Vancouver Province*, for allowing us to reprint these short clippings without specific acknowledgement. We have not edited these clippings, so at times their age is apparent (references to Bouey instead of Crow, for example). This should not affect the value of the clipping/question, nonetheless you may wish to alert your students to this matter.

Another strong feature of the book, which we had not fully realized until it was pointed out to us by pre-publication reviewers, is the quality of the chapter introductions. These introductions are concise overviews of the major ideas relevant to the topic at hand; they provide brief summaries of textbook material needed to tackle the questions in that chapter. One reviewer "would be willing to bet that students would pay just to have these short pieces even without the press clippings."

Some instructors have expressed concern that on the whole the questions we ask are difficult. This has been deliberate on our part. We feel that students learn far more if they are asked to address questions that push their understanding of economic principles further, rather than questions that require little thought to answer. This does not mean that easier questions do not have a role to play. They are valuable confidence-builders for the student and their presence allows the book to be used at the most elementary level. For these reasons we have ensured that a healthy percentage of the questions are of the "easy" type. But the majority of the questions are designed to challenge the student and could be classified by some instructors as difficult. But what may be a difficult question for one group of students may not be so for another group; the group's instructor is the best-qualified person to pass judgement on this. For this reason we have been reluctant to classify questions by level of difficulty; we strongly urge each instructor to review proposed student assignments carefully for difficulty (as well as for use of not-yet-studied concepts), utilizing for this purpose our suggested answers in the Appendix or in the instructor's manual. As a general rule, the earlier sections in both the micro and macro portions of the book are easier than the later sections, and within the short clippings of each section, the earlier clippings are easier than those appearing later.

As a further comment on the question of difficulty, consider the following four ways of asking students about the discouraged/encouraged worker phenomenon:

1. What are "discouraged workers"?
2. Explain what impact an increase in the number of "discouraged workers" would have on the measured level of unemployment.
3. An increase in the number of "discouraged workers" will
    a) lower the measured unemployment rate;
    b) leave the employment rate unchanged;
    c) lower the participation rate; or
    d) do all of the above.
4. "For example, Canada's jobless rate fell to 7% last month, the lowest number in two years. But if you think it's a sign that the economy is suddenly moving up, look again." How could a fall in the unemployment rate not be a sign that the economy is moving up?

The fourth question differs from the previous three in an important way. In each of the first three questions the student is told that the economic concept being examined is the "discouraged worker" phenomenon. In the fourth question no mention is made of discouraged workers; the student must identify the relevant economic concept on his/her own. In the words of a reviewer, "I am fond of telling my students that while it is important that they learn how to operate the various "tools" of the economist's trade, it is also important that they learn how to identify which tool to use in a given situation." Many students find this difficult to do; you may wish to make this fourth question easier by phrasing it as "Use the concept of discouraged workers to explain this clipping."

We advise instructors to read the section immediately below, directed to the student; it offers further comment on the nature of the book, along with advice to the student about how to develop answers to the questions posed in this book.

Preface to the Student

The purpose of this book is to give you an opportunity to apply to the real world what you have been learning in your textbook and in the classroom. This is accomplished by asking you to answer questions related to newspaper clippings, the most probable context in which you will be interacting with economics after your graduation. These clippings range from a few full-length articles to a multitude of "short clippings" consisting of only one or two sentences. For many students it is a big jump from textbook-type questions based on pure economic theory to questions based on real-world activity expressed in the cryptic, metaphor-filled style of news reporters. To ease this transition we offer some advice on how to address the questions in this book.

Each section in this book has an introduction providing a brief survey of economic concepts relevant to that section, and as part of the preface material we have provided a chart cross-referencing the sections in this book with most of the commonly-used principles textbooks. Before attempting any questions in a section you should read the introduction to the section as well as the relevant material in your textbook. Armed with this information you should be able to employ common sense and your knowledge of economic principles to generate answers to the questions found in that section.

Suppose, for example, that a clipping states "These high prices stimulated increased production" and you are asked "Does this statement refer to a shift in or a movement along the supply curve?" Since this question deals with the supply curve, you should review the theory of supply, paying particular attention to the distinction between a movement along and a shift in that curve. You should discover that a price change causes a movement along a supply or a demand curve, because price is measured on the vertical axis, and that any other change will cause a shift. This should allow you to deduce that in the context of the clipping the increased production results from a movement along the supply curve. Had the clipping stated "Lower costs stimulated increased production," the correct answer would have been a shift in the supply curve.

As a second example consider this clipping: "Theoretically, larger quantity drives the price down. The demand for premium Bordeaux is so great, though, that the rules of basic economics no longer apply. The 1970 vintage was the biggest in years and even veteran Bordeaux shippers assumed that prices would fall, but they rose dramatically." Suppose you are asked to comment on the statement that "the rules of basic economics no longer apply." Since this question addresses the issue of how new equilibrium prices are established in the market, a review of how supply-and-demand curve shifts lead to changes in price and quantity should be the first step. Such a review will reveal that increases in quantity, by shifting the supply curve to the right, should drive the price down, but that this result depends on the *ceteris paribus* assumption, i.e., assuming that nothing else changes. Thus it must be the case that something else has changed, and a look at a supply/demand diagram should make clear that to cause the price to rise in this example the demand curve must shift to the right by more than the supply curve does. That this is the case is hinted at in the clipping when it states that the "demand for premium Bordeaux is so great." Common sense suggests that the most probable source of the demand curve shift is the high quality of the 1970 vintage. If this is a relatively good year for Bordeaux quality, the demand for Bordeaux in normal years will not be a good guide to this year's demand.

Should you continue to have trouble with these questions, we advise that you exploit the Appendix which contains suggested answers to questions on selected full-length articles and all odd-numbered short clippings. By checking your answers against those given in this Appendix, you can reinforce your understanding of the economic concepts relevant to that section, check that you are structuring your answers appropriately, and pick up information useful in addressing other questions. When you do this, however, be sure to make an honest effort to generate your answer to the question before looking up our answer; in our experience, when students are given answers, they too often delude themselves into thinking that they could have produced similar answers.

It is our belief that by working through the contents of this book you will obtain a better understanding of economic theory and be able to view events in the real world more critically. And comments we have received from both students and parents indicate that also you should be able to conduct more interesting dinner-table discussions on economic issues!

Cross-References

The following table cross-references most of the sections of this book to specific pages in commonly used principles textbooks. By referring to this table, students needing review for any section should be quickly directed to relevant and easily available background reading. The following texts have been cross-referenced:

W.J. Baumol, A.S. Blinder, W.M. Scarth, *Economics: Principles and Policy*, 2nd Canadian edition (Toronto: Harcourt Brace Jovanovich, Canada, 1988).

A. Blomqvist, P. Wonnacott, and R. Wonnacott, *Economics*, 2nd Canadian edition (Toronto: McGraw-Hill Ryerson, 1987).

E.G. Dolan and R. Vogt, *Economics*, 3rd Canadian edition (Toronto: Holt, Rinehart and Winston of Canada, 1988).

R.G. Lipsey, D.D. Purvis, and P.O. Steiner, *Economics*, 6th edition (New York: Harper and Row, 1988).

A. MacMillan and B. Pazderka, *Microeconomics: The Canadian Context*, 3rd edition (Scarborough, Ont.: Prentice-Hall Canada, 1989).

A. MacMillan, *Macroeconomics: The Canadian Context*, 3rd edition (Scarborough, Ont.: Prentice-Hall Canada, 1989).

C.R. McConnell and W.H. Pope, *Economics*, 4th edition (Toronto: McGraw-Hill Ryerson, 1987).

P.A. Samuelson, W.D. Nordhaus and J. McCallum. *Economics*, 6th Canadian edition (Toronto: McGraw-Hill Ryerson, 1988).

D. Stager, *Economic Analysis and Canadian Policy*, 6th edition (Toronto: Butterworth and Co., 1988).

Cross References

Section		Baumol	Blomqvist	Dolan	Lipsey	MacMillan	McConnell	Samuelson	Stager
I-1	Supply and Demand	51-63	55-79	63-88, 445-49	57-75	25-61	47-63	55-68	33-50, 45-56
I-2	Elasticity	431-51	429-48	449-64	81-100	61-92	63-79	390-409	40-45, 58-60
I-3	Costs and Perfect Competition	453-522	467-500	495-53	176-237	107-50, 246-70	117-73	473-502, 412-17	394-431
I-4	Corn-Hog Cycle		436-42	661-75	113-19	65-70	307-27	418-21	510-14
I-5	Marketing Boards	562-63, 618-19	514, 544-46		119-22	212-13, 197-200	317-20	401-02	519-21
I-6	Government-Fixed Prices	63-69, 537-40, 740-42	69-72, 710-20	85-87, 667-68, 711-16	101-11, 123-27, 370-72, 346-49	30, 69-70, 37-39, 291-302	70-74, 261-78, 320-23	403-08, 227-28, 590-601	73-79, 515-18
I-7	At the Margin	411-21, 523-37, 621-24	449-60, 501-22, 645-59	467-79, 697	145-49, 160-68, 285-98	93-105, 227-29, 352-81	85-93, 352-53	422-35	371-80
I-8	Taxes and Subsidies	422-29, 433-35, 511-552, 575-92, 683-721, 761-71	95-101, 461-66, 501-22, 601-21, 645-59	483-93, 633-55, 670-71	132-44, 150-59, 472-518	99-102, 352-98	74-77, 96-102, 161-68, 329-37, 351-59	395-403, 435-444, 723-51	61-73, 79-84, 382-93
I-9	Foreign Competition	657-81	46-49, 306, 677-700	319-20	78-80, 394-436	320-44	668-89	521, 566-69, 840-53, 860-71	311-22
I-10	Imperfect Competition	555-72	549-71	579-628	257-83	181-213, 279-90	196-238	515-41	445-76

Cross References

Section		Baumol	Blomqvist	Dolan	Lipsey	MacMillan	McConnell	Samuelson	Stager
I-11	Regulated Monopolies	541-52, 615-36	523-47, 582-90	555-77, 655-61	240-57, 299-307	152-80, 362-66	174-95	515-30, 548-71	433-43, 490-97
I-12	Price Discrimination	645, 654-55	535-41	569-70	246-53	283-84	188-89	443-45	443-45
I-13	Government Franchises	751-54	770-76	730-32	336-43		287-91	498-500, 597-601	498-500, 597-601
II-1	Unemployment	92-100, 385-86	7-11, 151-63	6-7, 409-16	535-39, 555-58, 806-24	50-51, 264-88	277-79, 337-39, 418-28	213-34	89-118
II-2	Consumption and Aggregate Demand	145-96	185-210	166-78, 192-97	586-606, 703-06	115-22, 427-33	531-37, 551-59	122-35, 147-65	121-57
II-3	Investment and the Inventory Cycle	145-96	130-31, 153-54, 192-210, 285-95, 351-55	179, 281-86	613-27, 644-46, 652-67, 826-28	82-86, 180-82, 436-37	517-22, 547-49, 537-43, 595-97, 764-67	135-47, 191-210	131-57
II-4	Fiscal Policy	221-49	213-36	205-24	669-700	198-223, 345-48	574-84	168-83	205-35
II-5	Monetary Policy	251-317, 601-04	237-302, 341-45	227-316	708-71	139-76, 207-11, 344-52	443-98, 642-58	276-357	159-204, 271-82

Cross References

Section		Baumol	Blomqvist	Dolan	Lipsey	MacMillan	McConnell	Samuelson	Stager
II-6	Inflation	375-406	324-26, 362-72, 387-404	295-96, 385-93	774-95	126-28, 157-59, 278-80, 295-303, 310-12	428-39, 509-10	236-70	90-94
II-7	Interest Rates	104-05	262, 285-89, 390-99, 753-77	273-77, 295-96	358-62, 647-49, 736-44, 760-66, 865-66	126-31, 151-55	291-93, 434-35, 652-57	132-40, 304-12, 663-71, 678-80	170-72, 601-07
II-8	International Influences	319-54	305-28, 661-700	317-65	873-936	232-60, 384-93	663-725	168-90, 840-900	237-69, 307-42
II-9	Stagflation	375-406	357-85, 419-21	369-416	786-89, 797-805	264-312	611-41	252-70	205-13
II-10	Government Deficits	357-74	229-33, 295	218-25, 290-93, 302-03	688-700	205-206, 345-47	579-95	359-86	220-25
II-11	Free Trade	657-81, 688-92, 700, 779-82	688-93, 700, 779-82	322-30	406-38		665-93	840-53, 860-72	

xiii

Part I
MICROECONOMICS

Microeconomics *Sources of Press Clippings for Articles*

I-1A	*The Vancouver Sun*, June 18, 1985
I-1B	*The Vancouver Sun*, June 21, 1984
I-1C	*The Vancouver Sun*, July 8, 1988
I-1D	*The Toronto Star*, August 20, 1988
I-1E	*The Vancouver Sun*, July 5, 1988
I-2A	*The Vancouver Sun*, June 19, 1985
I-2B	*The Financial Post*, November 9, 1985
I-2C	*The Vancouver Sun*, March 19, 1980
I-3A	*The Vancouver Sun*, May 17, 1985
I-3B	*The Vancouver Sun*, March 31, 1988
I-4A	*The Financial Post*, May 7, 1988
I-4B	*The Financial Post*, March 22, 1980
I-5A	*The Vancouver Sun*, June 21, 1983
I-5B	*The Toronto Star*, April 16, 1985
I-6A	*The Financial Post*, August 17, 1985
I-6B	*The Toronto Star*, April 15, 1988
I-7A	*The Financial Post*, September 12, 1988
I-7B	*The Globe and Mail*, Toronto, April 10, 1976
I-8A	*The Vancouver Sun*, June 5, 1981
I-8B	*The Financial Post*, December 7, 1987
I-8C	*The Vancouver Sun*, May 11, 1985
I-9A	*The Vancouver Sun*, October 18, 1986
I-9B	*The Globe and Mail*, Toronto, October 11, 1988
I-9C	*The Vancouver Sun*, June 8, 1985
I-10A	*The Vancouver Sun*, June 18, 1988
I-10B	*The Vancouver Sun*, January 29, 1977
I-11A	*The Province*, Vancouver, April 28, 1976
I-11B	*The Vancouver Sun*, October 6, 1984
I-12A	*The Globe and Mail*, Toronto, August 13, 1976
I-12B	*The Toronto Star*, March 6, 1976
I-13A	*The Globe and Mail*, Toronto, May 14, 1976
I-13B	*The Financial Post*, August 17, 1985

I-1 Supply and Demand

Microeconomics uses supply-and-demand curves as its basic tools of analysis. Before students can apply microeconomics to business decisions or to policy problems, they must be skilled at interpreting and using these curves. This first section aims to develop this ability by looking at the supply-and-demand curve interpretation of some markets which are close to being "perfectly competitive": There is a large number of buyers, a large number of sellers, and a homogeneous product.

Equilibrium in these markets (a price and quantity combination that satisfies both demanders and suppliers) is usually quickly established through price changes, keeping the market at the intersection of its supply-and-demand curves. Most of the economic analysis surrounding these markets, therefore, is centered on shifts in these curves. Supply-and-demand curves are derived on the *ceteris paribus* principle: They trace the supply of or demand for something as its price changes, *other things remaining the same*. The demand for television sets, for example, depends on their price, but it also depends on the overall level of income in the economy. As the price changes we move up or down the demand curve; as the level of income changes the entire demand curve shifts to the right or the left.

These shifts in the demand or the supply curves throw the market out of equilibrium, instigating a disequilibrium reaction on the part of the market participants. Both price and quantity adjust, moving the market (in a stable system) to the new equilibrium position.

I-1A Oil

Oil supply key to natural gas prices, confab told

JASPER, Alta. (CP)—Oil will continue to be the dominant primary energy source well into the 21st century, Shell Canada president Jack MacLeod said Monday.

"The specific quantities of oil available and the pricing of oil will, therefore, continue to have a major influence on the pricing of natural gas over the same, rather indeterminate period," MacLeod told about 450 delegates attending the Canadian Gas Association's annual conference here.

He predicted the key players of the Organization of Petroleum Exporting Countries will remain sufficiently cohesive to hold oil prices to a relatively modest rate of decline for the remainder of the 1980s.

"After that, market forces will support a turnaround to a modest rate of increase in real terms throughout the 1990s."

MacLeod said there are sufficient reserves of natural gas throughout the world to provide adequate supply until the end of the 20th century.

But there may be national or regional imbalances, he said, and the challenge to the gas industry and to country governments will be to bring remote, expensive sources to market at market value.

"Buyers' markets prevail and are likely to, throughout most, if not all, of the rest of this century," he said.

Note: Suggested answers to the questions below appear in the Appendix.

1. Explain the second paragraph by referring to supply-and-demand diagrams for oil and natural gas.
2. In terms of these supply-and-demand diagrams, in what sense will the "key players" have to be "cohesive to hold oil prices to a relatively modest rate of decline"?
3. What market forces do you think are referred to in the fourth paragraph?
4. What is meant by "increase in real terms"? (fourth paragraph)
5. What is meant by market value? (sixth paragraph)
6. What is a "buyers' market"?

Newsprint use climbs, MB to increase price

With newsprint markets rebounding, MacMillan Bloedel Ltd. said Wednesday it will increase the price of newsprint in Canada and the United States.

The price in the U.S. goes to $535 (U.S.) a metric tonne from $500 (U.S.) a metric tonne. In Canada, the price will increase to $585 a metric tonne from $540 a metric tonne. Both increases are effective Oct. 1, 1984.

A contract-performance incentive, which gives a five-per-cent rebate to U.S. customers who buy 100 per cent of their annual contracted tonnage, will remain in effect.

Eric Lauritzen, vice-president of pulp and paper marketing at MB, said: "The markets in the U.S. and Canada are showing signs of a dramatic recovery at the moment. This is true for both daily newspapers and commercial printers."

He said the increased consumption is a reflection of the "very buoyant U.S. economy." The U.S. newsprint market accounts for two-thirds of the output at MB's Port Alberni and Powell River mills.

"We forecast by the fourth quarter of this year that supply and demand of newsprint will be in balance for the first time in two years," said Lauritzen.

"And by the end of this year, we have an excellent chance of operating at full capacity for the first time in 18 months."

Lauritzen predicted that "supply and demand in the U.S. will be in balance through the end of 1986."

He said the recent three-year settlement with the Canadian Paperworkers Union and the Pulp, Paper, and Woodworkers of Canada will allow MB to guarantee continuity of supply.

Note: Suggested answers to the questions below appear in the Appendix.

1. What is causing the price of newsprint to increase in North America? Use a supply-and-demand diagram in your answer.

2. How does the contract-performance incentive affect the demand for newsprint? Show this on a demand curve.

3. By referring to the article explain how the newsprint industry is responding to the increased demand, i.e. a movement along or shift of the industry supply curve?

4. Does Lauritzen's interpretation of supply and demand being in balance differ from an economist's? Explain. (paragraphs six to eight)

I-1C Pulp

Shortages feared if pulp negotiation results in a strike

By VALERIE CASSELTON

Pulp talks have turned the eyes of world pulp consumers to B.C. in the fear a work disruption would upset supply and send pulp prices soaring further, analysts and consumers said Thursday.

"If we had a strike here, my God, the price of pulp would skyrocket," said forest industry analyst Charles Widman, president of Widman Management Ltd.

B.C. produces about 15 per cent of the world's pulp and demand is so great that loss of that supply would significantly affect world supply and prices.

"There never has been, in the past 20 years, a more profitable period for pulp," said Widman. "In the last $2^1/_2$ years they have raised the price every quarter. There's hardly any commodity I know of that you can increase the price every three months and the market will accept it."

In the U.S., prices have increased quarterly despite consumers' protests, said Dave Ramm, West Coast sales manager for Hollingsworth and Vose, a Massachussets firm specializing in filter products for automotive, rail and hydraulic use.

"The pulp industry is running at 100 per cent capacity, so any shortage could have really serious consequences," Ramm said from his office near San Francisco. "People are nervous that if the pulp industry is running at full capacity, if 10 of 15 per cent is pulled off the line, there would be an interesting situation."

Steve Sutta, a California waste paper dealer, said he stands to make a profit from any B.C. pulp strike because Asian pulp consumers already are trying to line up alternative supplies.

Ramm said Canadian unions "would be crazy not to ask for a huge raise" but he said a sliding pay scale that would move lower if the boom passed would be also be sensible.

Ramm said his company is an "insignificant" consumer of pulp compared to newspapers, which consume massive volumes of pulp for newsprint.

---

1. Use a supply-and-demand diagram to explain how a strike will cause the price of pulp and paper to rise.
2. What must have been happening in this diagram over the last two and a half years for the market to accept price increases every three months? (fourth paragraph) What could be causing this?
3. Use a supply-and-demand diagram to explain how the rising price of paper affects the filter industry.
4. If the pulp industry is running at 100 per cent capacity, what can we say about the shape of the supply curve at the current output level? Explain.
5. Use a supply-and-demand diagram to explain how the waste paper industry is affected by rising pulp prices.
6. Explain why a sliding pay scale would be sensible (second-last paragraph).

I-1D Antifreeze

Ethylene shortage sending costs for antifreeze through the hood

LONDON, ONT. (CP)—A world-wide shortage of ethylene, used to make many types of plastic products, is boosting prices for such consumer products as antifreeze and garbage bags.

Since late 1987, the price for ethylene has increased to 67 cents a kilogram from 29 cents—a jump of 131 per cent.

"Our costs have essentially doubled and will more than double by the end of the year," said Gord Robertson, vice-president of marketing for the automotive products division of First Brands (Canada) Corp.

Among other products, the company produces Glad garbage bags and Prestone antifreeze.

Robertson said Prestone, which could be picked up on special for $6.99 for a 4.5-litre jug last winter, will retail no lower than $9.99 this season. At small outlets, like service stations, the price could exceed $12, he said. Prices are also being increased for the company's Glad garbage bags.

In the United States, dairies are already using glass and cardboard containers in place of plastic jugs to cut costs.

Ethylene producers are responding to the need.

Polysar Ltd., based in Sarnia, recently announced plans to spend up to $250 million to increase production of the chemical.

Polysar, Nova Corp. and Dupont Canada are the major Canadian users of ethylene. Nova Corp.'s acquisition of Polysar in June gave it control of about 80 per cent of ethylene production in central and western Canada.

"There's a worldwide shortage because of the rationalization of the petrochemical industry after the recession," said Firman Bentley of Polysar.

At the time, sagging demand and low ethylene prices caused some producers to trim back production and others to shut down completely. But a booming world economy has fueled the demand for plastic.

The supply-demand pinch was made worse by the explosion at the Shell Oil plant in Norco, La., and a fire at Texaco's plant in Port Arthur, Texas.

---

1. To an economist, what does the word "shortage" mean, as used in the first paragraph?
2. Use a supply-and-demand diagram to explain how the world-wide shortage of ethylene is boosting the prices of antifreeze and garbage bags.
3. Use a supply-and-demand diagram to explain how the increase in ethylene prices is affecting the glass and cardboard container industry. (sixth paragraph)
4. The second-last paragraph states that producers are reducing supply: explain whether this is a shift in the supply curve, a movement along it, or both.
5. a) What is a "supply-demand pinch," referred to in the last paragraph?
   b) What has caused it, and how would you show this cause (or causes) on a supply-demand diagram?
   c) Using your diagram, explain how the explosion and the fire have made this pinch worse.

I-1E Beer

Drought threatens beer prices

Beer prices may rise slightly in the future because of a Prairie drought that has dramatically increased barley prices, a spokesman for Labatt Breweries in Vancouver said today.

Philip Carter was commenting on a report from Toronto that quoted a Labatt's spokesman there as saying beer drinkers could expect to see prices rise by not more than 50 cents for a case of 12.

Carter said rain on the Prairies could change all predictions. "We will obviously try to hold prices down as long as possible. It is in no one's interest to see the price of beer rise," he said.

---

1. Explain how a drought affects the barley market to cause prices to rise.
2. Use a supply-and-demand diagram to explain how an increase in barley prices can affect the beer market to cause beer prices to rise.
3. From the perspective of the beer industry, why is it not in their interests to have the price of barley go up? Can't they just pass this increased cost on to consumers?

Short Clippings

Note: Suggested answers for the odd-numbered questions below appear in the Appendix.

A. Interpretation

A freeze in southwestern U.S. and Mexican fields has resulted in damaged crops and inflated prices for imported produce such as lettuce, broccoli and cauliflower.	1. How would this be interpreted on a supply-and-demand diagram?
Gagne said that, for example, higher prices in the U.S. might make it possible for some American producers who closed during the last two years to reopen.	2. How would this be interpreted on a supply-and-demand diagram?
Spurred by a strong economy and scarcity of supplies, Canada's pulp and paper producers. . . will celebrate the holiday weekend by sending the prices of their products soaring to record highs.	3. How would an economist refer to "scarcity of supplies"?
And for Canadian wineries, which haven't yet convinced Canadians to choose their products over imports even with markups, it's going to be a struggle. As Bright's Diston puts it, "There's too much wine in the world."	4. How would an economist interpret the phrase "There's too much wine in the world"?
Skyrocketing property values in Vancouver's west side can be blamed on greedy vendors and not foreign investors, says a Vancouver real estate agent who specializes in the district. A classic example, she said, was one client who owned property in the west side valued $500,000. "The man wanted $750,000 because he knew some Chinese buyer will come along and take it," she said. "Now, you tell me who's driving up the market."	5. Do you agree with this real estate agent? Why or why not?

"Low interest rates and a strong economy create a natural demand (for housing) and the demand started very quickly this year—by the third week of January," Jackman said. "Everyone seemed to jump into the market all at once and we didn't have enough product to go around."

6. a) Show the state of the housing market on a supply-and-demand diagram.
   b) How would an economist describe not having enough product to go around?
   c) How will the market ensure there is enough product to go around?

On suites renting from $400 to $500 a month, a landlord recently cut the rents by $25 to $50 a month to accommodate his cash-strapped tenants. That's an exceptional case, he noted, but it still reflects market conditions for many landlords.

7. What market condition is being reflected here? Refer to a supply-and-demand diagram.

"There are still 2,500 vacant units in the Lower Mainland and just so many tenants to go around. Landlords still have to take measures to get people in their buildings."

8. Explain, by using a supply-and-demand diagram, the current position of the market and what measures the landlords will have to take.

And in the Lower Mainland, according to the Rental Housing Council of B.C., landlords are continuing to cut rents or at least are delaying increases, despite a falling vacancy rate.

9. Using a supply-and-demand diagram explain the state of the rental market. Explain how the falling vacancy rate could be consistent with rent cuts.

"The mills are booked two months solid," said Paul Pannozzo, a lumber purchaser for the Lumberland retail chain. He said what happens then is that buyers are forced to bid higher prices in order to snatch enough material for their needs, before it's sold down south.

10. Explain what is happening here via a supply-and-demand diagram.

"There's resistance at the retail level. The wild mink production in a year would be around a half million and there's maybe 35 million ranch mink. They're over-produced. They should have improved their product and decreased production."

11. a) Use a supply-and-demand diagram to illustrate the current state of the mink market.
    b) What is meant by resistance at the retail level? How will the market eliminate this resistance?

Bill Brown, general manager of Andres Wines of Port Moody, said Wednesday that he expects an amicable solution will be reached that will move the industry towards reflecting consumer preference for white wines.

Andres cancelled contracts with 16 red grape growers last May, Calona with 62 growers and Casabello with one grower because they say there is already a five-million-gallon surplus of red wine.

12. a) How is "consumer preference" affecting the market for red and white wines? Use a supply-and-demand diagram for each market.
    b) What adjustments will occur in the wine industry to reflect the current problem? In your answer refer to a supply-and-demand diagram for each wine.

B. Shifting Supply And Demand Curves

Nickel has benefited from the same basic economic principles that have already spurred copper to its highest levels in years: while demand has been unexpectedly buoyant, supplies have been surprisingly tight.

1. Interpret this clipping in terms of a supply-and-demand diagram.

Last fall, hog prices peaked at more than 90 cents a pound in Toronto and 60 cents in Chicago. These high prices stimulated increased production and marketing levels now reflect this increase.

2. Does the increase in production reflect a movement along or a shift of the supply curve? Explain.

Lettuce, which usually sells for 89 to 99 cents a head at this time of year, is currently selling at a gourmet price of up to $2.

The price hike can be chalked up to bad weather in California: "It's been a dog's breakfast. They've had hot, cold, too wet, too dry. Put them all together and we've got this beautiful mess," says John Christiansen, Canada Safeway's produce merchandising manager.

3. Use a supply-and-demand diagram to illustrate the influence of California weather on the price of lettuce.

The performance of real estate stocks depends largely on one key economic factor, interest rates. Lower rates, of course, stimulate higher property values, while keeping down finance costs for both construction and mortgages.

4. Use a supply-and-demand diagram to explain how lower interest rates stimulate higher property values.

Canadian mills are now increasing production as U.S. housing starts rose dramatically in June. Lumber prices are getting much better now and some mills are adding new shifts as lumber inventories in the U.S. have dropped to historic lows.

5. a) Does the increase in U.S. housing starts cause a movement along the Canadian demand curve for lumber or a shift in this curve? Explain.
   b) Are the added shifts a reflection of a movement along the supply curve or a shift in the supply curve? Explain.

Our construction industry has been encouraged by the proposed removal of rent controls in some provinces and by the continued improvement in the efficiency of our lumber mills.

6. a) Would the removal of rent controls imply a movement along the supply curve for housing or a shift in that curve? Explain.
   b) Would improvements in the efficiency of our lumber mills imply a movement along the supply curve for housing or a shift in that curve? Explain.

Inflation rates are low, interest rates are stable, the economy is stable and inmigration levels are lower. Toronto-based housing analyst Frank Clayton said those factors should ensure a less dramatic runup or falloff in prices as long as speculators don't fuel the market.

7. Explain why low inflation rates, stable interest rates, a stable economy and lower immigration levels should ensure a less dramatic runup or falloff in prices.

Export sales of electricity to the United States jumped to $142 million from $11 million during the previous nine-month period.
The increase in U.S. sales was caused by low water conditions in the Pacific Northwest and increased fossil fuel costs in California.

8. Using a supply-and-demand diagram, explain how the following could increase sales of electricity.
   a) low water conditions
   b) increased fossil fuel costs

Improved grain production methods and increased acreage have more than offset population increases.

9. By referring to a supply-and-demand diagram, explain what is happening in the grain market.

The exchange rate, giving U.S. lumber buyers a better deal for their dollar, hasn't helped any as far as Canadian buyers are concerned.

10. How does the exchange rate hurt Canadian buyers of lumber in this clipping?

. . . he said the pricing story could change completely if B.C. runs into labor problems, since there are no new contracts yet in the forest industry.

11. What have B.C. labor problems got to do with the price of lumber?

I-1 *Supply and Demand* 11

Higher prices for feed and the interest on loans result in increased costs that are one way or another passed on to the consumer.	12. a) How will the free market pass these increasee costs along? Refer to a supply-and-demand diagram. b) Does this mean the price to the consumer will rise by exactly the jump in cost? Explain why or why not, using a supply-and-demand diagram.

C. Substitution And Demand Changes

Consumers who turn down beef and pork, however, are faced with rising prices for substitute protein products such as eggs, cheese, and poultry.	1. Explain why consumers refusing to buy beef and pork because of their high prices are likely to find "rising prices for substitute protein products." Refer to a supply-and-demand diagram.
Consumers have found that if they grind the coffee finer, they can get more cups from a package. Others are drinking more tea.	2. The price of coffee has risen sharply due to a drop in coffee supplies. How will the price of tea be affected? Explain.
The expected price drop for cherries due to the bumper crop this year won't be as much as we thought, due to short supplies of pears and peaches.	3. Explain this clipping by referring to a supply-and-demand diagram.
The availability and prices of beef substitutes such as pork, poultry and fish influence beef prices. In recent months, farmers in Canada and the United States have increased hog production, causing pork prices to fall and contributing to the slight decline in beef prices.	4. a) Explain how beef prices are influenced by the availability and prices of beef substitutes. b) Interpret the second sentence of this clipping in terms of a supply-and-demand diagram.
Pork will be more competitive to beef and all red meat prices will experience downward pressure.	5. What is happening to the price of pork? How will "all red meat" experience downward pressure? Explain via a supply-and-demand diagram.
Low pork supplies resulted in beef being very competitively priced in relation to pork.	6. Explain the rationale behind this statement.

Dairy herd culling has increased beef supplies and forced prices lower. Now, increased hog production will provide heavier competition to beef sales.	7. a) What is happening to cause the price of beef to fall? Is it a movement along or a shift of the supply curve? b) How will increased hog production "provide heavier competition to beef sales"?
There is no question that the phenomenal prices these wines bring have a tendency to pull the prices of most other wines along with them. Six dollars is an extraordinary price to pay for a bottle of Chianti—until one compares it with a $12 Bordeaux.	8. Give the the economic explanation for the tendency reported here. Refer to supply-and-demand curves.
The rising cost of building material for new homes has helped increase prices for older homes.	9. Using supply-and-demand diagrams, illustrate why the price of older homes is increasing.
The low vacancy rate for apartments has placed an additional burden on those wanting to buy a home.	10. Using supply-and-demand diagrams, explain what this additional burden is and how it is caused.

D. Connected Markets

Even though there is no shortage of eggs now the consumer can expect to pay more in the future as a world shortage of grain used for feed is expected.	1. a) What is implied about the state of the egg market? b) How and why would a world storage of grain lead to higher egg prices? Explain using a supply-and-demand diagram.
Bumper grain crops this year in both Canada and the U.S. will further encourage meat production.	2. a) What has happened in the grain market? b) Explain the impact of the bumper grain crops on the supply-and-demand diagram for meat.
The drought has sent feed prices soaring, and producers are reluctant to invest in breeding stock. Hog farmers are selling off their stock at plummeting prices, and many of them will lose money this year.	3. Explain this clipping by showing what is happening on supply-and-demand diagrams for feed and for pork.

Wheat futures prices bounded higher amid increasing fears that disease and drought could significantly reduce yields from the rapidly maturing winter wheat crop.

The early runup in grain prices had a strong negative impact on cattle futures on the Chicago Mercantile Exchange.

Higher grain prices would mean higher feed prices and narrower profit margins for cattle producers, which could ultimately lead to higher beef prices and a corresponding drop in demand for beef, analysts said.

4. a) Use a supply-and-demand diagram to explain how disease and drought could cause a "runup in grain prices."
b) Use a supply-and-demand diagram to explain how this rise in grain prices could lead to higher beef prices.
c) Is the drop in the demand for beef, noted at the end of the clipping, a shift in a demand curve or a movement along it?

Demand for the new product has so taxed the new mill that, according to one Vancouver retailer, there's a wait of 30 to 60 days now to get orders filled.

Oriented strand board is a waferboard-type panel that meets most of the strength and durability requirements of plywood, and sells consistently at 10 per cent less than the spruce plywood it competes with.

The manufacturing process takes almost any tree species (aspen is the most common) and cuts it into strands about the size of a band-aid. The strands are placed (oriented) perpendicular to each other in three layers, and glued with waterproof glue.

The result is a panel that is, for most uses, the equal of plywood at a consistently lower cost.

5. a) What effect would the introduction of the new cheaper board have on (i) the plywood market and (ii) the housing market? Explain using a supply-and-demand diagram for each market.
b) What does the first paragraph indicate about supply relative to demand for the new product? What would you predict will happen to the price of the product (i) in the short run, and (ii) in the long run?

Like Lloyd Quantz of Agri-Trends Research Inc. in Calgary, Brinkman says saturated world grain markets and the prospect of lower livestock prices in 1988 will continue to depress farm values over at least another two years.

6. a) Use a supply-and-demand diagram to explain what is happening in the grain market.
b) How will events in the grain market affect the livestock market?
c) Why will farm values be depressed?

E. Miscellanea

"If copper stayed at $1.25 a pound for two or three months, that could cause some inflationary pressures—but the price won't stick."

"The system is more efficient than it used to be. If higher copper prices endured, they would be self-correcting. More production would come on line and there would be more substitution."

1. Use a supply-and-demand analysis to explain this clipping. Hint: Focus on the difference between short-run and long-run curves.

Canadians spend more hours talking on the phone than anyone else in the world. That's a fact B.C. Tel's Roger Varley readily admits. Tel's talk of "savings" for 60 per cent of subscribers is based on a factor called "repression"—the already proven tendency for consumers suddenly slapped with LMS to curtail their telephone usage. When every call costs money, you make fewer calls.

But "repression" isn't permanent. B.C. Tel data suggest that after initial repression takes place, telephone usage will rebound by seven per cent a year, so that in 10 years time the effect of repression will be gone.

2. Use supply-demand analysis to explain this clipping. Hint: Focus on the difference between short-run and long-run curves. Note: LMS stands for locally-measured service, a system in which every local phone call costs money, as opposed to a flat-rate for unlimited calls.

McCance said the association acknowledged that the warning (of future lumber price increases) has the potential of being a self-fulfilling prophecy.

3. Using a supply-and-demand diagram, explain how a warning of future lumber price rises could be a self-fulfilling prophecy.

Perkins says the dry weather has already affected grain futures markets overseas and other buyers continued to bid up fall delivery prices on world markets last week.

4. Dry weather would seem primarily to affect *supply*; and yet it appears that the "bidding up" of grain futures is the result of extra *demand*. Explain with the help of a supply-and-demand diagram.

If nickel prices keep increasing, customers will start turning to other materials, says the chairman of Inco.

There is a danger of prices going too high and even if they remain at the same level since late last year, there could be a permanent decrease in nickel demand, Donald Phillips told the company's annual meeting Wednesday.

5. Explain how a permanent decrease in market demand could occur.

The irony behind these prices is that there is no shortage of wine. The famous names of Bordeaux and Burgundy make up less than one-fifth of one per cent of the world's annual output.

6. The clipping refers to sharply rising prices of well-known Bordeaux and Burgundy wines. Remove the irony by explaining the meaning of "shortage" in this context.

Theoretically, larger quantity drives the price down. The demand for premium Bordeaux is so great, though, that the rules of basic economics no longer apply. The 1970 vintage was the biggest in years and even veteran Bordeaux shippers assumed that prices would fall, but they rose dramatically.

7. Is it true that "the rules of basic economics no longer apply"? Explain your answer by means of a supply-and-demand diagram.

The same thing happened in July and August when quota cheaters drove the spot price of oil down sharply after its brief rally beyond $20.

8. Illustrate this on a supply-demand diagram.

The current slump in the world oil market, characterized by volatile and low oil prices, is setting fundamental forces in motion which will combine to create another oil crisis. No matter how you look at it, it is just a matter of time before crude oil prices begin to increase in real terms and, in general, the longer that world oil prices stay low, the faster they will rise in the future and the larger the increase in price.

9. Explain the nature of the fundamental forces referred to. Hint: Consider the effect on the demand for oil.

The controversial 250-page document, prepared for the federal consumer affairs department, was made public this week. It calls on the government to shatter the long, status-laden tradition allowing professionals such as doctors and lawyers to set their own working rules. . . . they alone can license anyone wishing to practice. The report finds this the most distasteful aspect of self-regulation.

"Occupational licensing has typically brought higher status for the producer of services at the price of higher costs to the consumer."

10. How would you explain the essence of this item in terms of supply-and-demand curves?

The future of family farming may not lie on the farm at all, Henry Compton stresses. "Canadian consumers are getting a terrific deal. They must expect to pay more in future if farmers are going to stay in agriculture."

11. a) What logic lies behind the last sentence?
    b) Why hasn't this same logic led to higher food prices over the past 50 years as the number of farmers has shrunk?

I-2 Elasticity

The price elasticity of demand is the percentage change in demand caused by a one per cent change in price. It is a mechanical property of demand curves that tells us something about how sensitive demand is to price changes. Given magnitudes of a price change and the corresponding quantity change, elasticity can easily be calculated with a little arithmetic: Divide the percentage change in quantity by the percentage change in price. (The resulting negative sign is often ignored.) An inelastic demand curve is one with low elasticity: A price change elicits very little change in demand. An elastic demand curve is one with high elasticity: A price change alters demand considerably. Note, however, that elasticity really refers to a *point* on the demand curve, not the entire demand curve itself: Different points on the same demand curve can and usually do have different elasticities associated with them.

The importance of all this relates to the impact on total revenue of a price change. In the inelastic case, a small percentage price rise causes an even smaller percentage decrease in quantity demanded, so that total revenue, the product of price and quantity, rises. In the elastic case, a small percentage price rise causes a large percentage decrease in quantity demanded; total revenue falls. When elasticity is exactly unity (an increase in price causes the same *percentage* decrease in quantity demanded) total revenue remains unchanged. If demand is inelastic, and thus total revenue rises when prices rise, there is a great temptation to businessmen to raise prices to increase profits. Only competition or government regulation can protect the consumer from being exploited in such markets.

Elasticity of demand (or of supply) can be important for two other reasons. In analyzing the dynamics of any market (how it moves from one equilibrium to another), elasticities are important in determining whether or not cycles are generated. For example, a little excess supply might cause a disruptively large fall in price if demand is inelastic. In applying government policy, the success of that policy in terms of its stated goals may very well depend on the relevant elasticities. For example, a lower exchange rate may not stimulate exports if the demand for our exports is inelastic. Both these roles of elasticity should become apparent in later sections of this book.

There are two major problems in measuring elasticity; both appear in the articles in this section. First, elasticity is a *ceteris paribus* concept—it relates to the quantity change resulting from a price change *with everything else held constant*. In the real world, however, we seldom have the opportunity of observing a situation in which price is the only element affecting demand that has changed. Casual elasticity measures, such as the ones that can be calculated from newspaper articles, should therefore be treated as ballpark figures. The second major problem is that elasticity is supposed to be calculated using very small changes in price and quantity, because elasticity may change as we move from point to point along the demand curve. If we have numbers only for large changes in price and quantity, the resultant calculated elasticity will not accurately represent the elasticity at either the old price or the new price. For example, the difference between prices of $10 and $15 could be conceptualized as either a 50 per cent rise or a 33 per cent fall. Elasticity estimates are usually therefore computed (if possible) using averages of the old and new prices and quantities as the bases from which percentages are calculated, and the elasticity is related to this average position. This measure is called the arc elasticity.

I-2A Natural Gas

Gas sales rise, but revenue slips

TORONTO (CP)—Canadian natural gas companies sold six per cent more gas to the United States last year than in 1983, but lower prices brought 0.5 per cent less revenue into Canada.

Canadian gas companies shipped 21.4 billion cubic metres to the U.S. last year, up from 20.1 billion in 1983, according to figures released Tuesday by the Canadian Gas Association.

But lower prices permitted under new federal regulations instituted last year reduced the over-all take from $3.94 billion to $3.92 billion.

Note: Suggested answers to the questions below appear in the Appendix.

1. Calculate the elasticity of demand for natural gas sold to the U.S.
2. The calculation made in question 1 assumes that the demand curve in the U.S. has not shifted. Explain in what way your calculation would be biased if the demand curve had shifted.
3. Why would the natural gas industry lower prices if they made less money?

I-2B Software

Sales soar following package's price cut

A brave switch in marketing strategy appears to be paying off for Bedford Software Ltd. of Burnaby, B.C., which has slashed the retail price of its well-accepted accounting package to $149 from $1,200, and has seen orders soar.

Bedford began selling the accounting package in January to enthusiastic reviews. Thomas G. O'Flaherty, vice-president, marketing, says the $1,200 price was chosen for what he terms "historic reasons"—simply, it was the sort of price commonly charged in the industry.

Sales were perking along quite nicely at about two a day when the company decided to rethink its entire marketing strategy.

It reasoned that while a $1,200 price tag might be acceptable to a large company purchasing a top-of-the-line micro with all the bells and whistles, it could be too rich for the owner of a small business—the type of person for which the package is intended.

"Our sales units are more than 20–30 times what they were before the price decrease, and our revenues are up four times," O'Flaherty says.

---

1. Given the fact that revenue has increased four times, what can you conclude about the elasticity of demand for this accounting package?
2. Again considering the increase in revenue, find the actual increase in sales and calculate the arc elasticity of demand.

Gasoline: does price affect consumption?

MICHAEL VALPY IN OTTAWA

Today an unsettling issue: the relationship between price and consumption of gasoline. This is no definitive essay on the subject, but it is a look at some of the assumptions.

John Crosbie's budget estimated that the increase in the gasoline excise tax of four cents a litre (16.5 per cent at the pumps) "could amount" to a reduction in gasoline consumption in Canada of almost three per cent in a year.

Jimmy Carter, in his anti-inflation statement last Friday, estimated that his "gasoline conservation fee" of 2.5 cents a litre (5.7 per cent at the pumps) "should reduce" oil imports by as much as 36.5 million barrels in a year. Or, according to a commonly used formula, reduce gasoline consumption by perhaps one per cent over the year.

You will notice the heavy qualifications.

Prime Minister Pierre Trudeau, in the one press conference he gave during the election campaign, debunked the idea of using price as a conservation measure. "The course of using prices in order to deter people from using electrical apparatuses or gasoline or diesel fuel hasn't worked, doesn't appear to be working," he said. "I'm told that people are buying just as many big cars as before.

"I think most economists will tell you that there's, on the contrary, a substitution effect. If it costs more to drive your car, you don't stop driving your car. You perhaps don't buy that extra fishing rod, or you don't take that second trip to Miami at Easter. But you don't put your car in the garage and say, I'm leaving it at home. The motorcar is too ingrained in the psyche of North American males and females to think that they are going to cut part of their budget they spend on that before they cut a lot of other things."

Former prime minister Joe Clark does believe that conservation is affected by price.

"I am persuaded there is a clear relation," he said in a Sun interview during the election. "The dramatic kind is experienced with automobile purchases. When the OPEC price increases came, there was a 40 per cent shift, a 40 per cent increase, in the purchase of compact cars. There has been other evidence that is more detailed—and in some aspects less conclusive—that the department of energy, mines, and resources has. . . ."

Let's look at the evidence of the energy, mines, and resources ministry.

EMR's economic and policy analysis branch uses the formula that a 10 per cent real increase—that is, after inflation—in the price of gasoline will result in a reduction in consumption by 1.5 - 2 per cent in a year. They say the formula is 90 to 95 per cent accurate, based on evidence from Canada, the United States, and Europe.

For Canada, they offer evidence that in the period 1963 to 1977 gasoline consumption per automobile has been lowest when prices are highest. For example, in 1973 when gasoline was 76.5 cents a gallon (the cheapest price, in 1977 dollars, over the 15 years), consumption was at about 880-plus gallons per car. In 1977, when gasoline cost nearly 87 cents a gallon (the highest price), consumption was down to 800 gallons per car.

Keep in mind, of course, that while the EMR studies show per-auto consumption of gasoline dropping in relation to price, total consumption in Canada increased by 4.3 per cent in the first nine months of 1979. Thus while each of us may be using somewhat less, more of us are using it over-all.

But in America total consumption over the same period dropped by four per cent. The American Petroleum Institute believes that is a direct result of sharply increased prices. As further evidence, the Americans point to the fact that in 1979, with fears of gasoline shortages, subcompact cars that were more energy-efficient accounted for 62 per cent of new-car sales.

In Canada, on the other hand, with lower gasoline prices, the subcompacts accounted for only 47 per cent of new-car sales in 1979—a drop from 48.5 per cent in 1978. That lends weight to Mr. Trudeau's statement that people are buying as many large cars as ever.

At the same time, there is evidence of the American trend taking hold here. In October, import-car sales rose by 11.3 per cent (over October 1978), in November they were up 15 per cent, in December up 72.6 per cent, in January up 61.5 per cent.

1. Calculate the price elasticity of the demand for gasoline according to (a) Crosbie, and (b) Carter.
2. What does Trudeau think this price elasticity is?
3. Is the substitution effect noted by Trudeau the same as the substitution effect in economic textbooks when changes in demand due to price changes are discussed? Explain why or why not.
4. Is the substitution effect noted by Trudeau stronger or weaker when the demand for gasoline is more inelastic? Explain.
5. Explain why the EMR analysis stresses real increases in the price of gasoline.
6. Calculate the EMR real-price elasticity of demand for gasoline as given by their formula which they claim is 90 to 95 per cent accurate. Comparing this number to those of Crosbie and Carter, would you guess that Crosbie and Carter are implicitly talking of real price increases or just unadjusted price increases for gasoline? Explain your reasoning.
7. It's not clear from the article if the numbers used to calculate elasticities refer to per-automobile or total consumption. Assuming the latter is the case for the Crosbie and Carter numbers, do your calculated elasticities in question 1 underestimate or overestimate the elasticity of per-automobile demand for gasoline? Explain. Is your answer consistent with the numbers given by EMR for the 1973 to 1977 period?
8. Using information supplied in this article, would you say that the short-run per-automobile real-price elasticity underestimates or overestimates the long-run elasticity? Explain.

Short Clippings

Note: Suggested answers for the odd-numbered questions below appear in the Appendix.

A. Elastic vs Inelastic

. . .the other part of the bullet the government has not even touched its gums to is conservation. There is probably not a respectable economist or planner in the world who believes price is a factor in conservation. People will gas up their cars with the same regularity and keep their thermostats at 22 degrees whether oil prices go up $4 a barrel or $40.

1. What does the author believe about the price elasticity of the demand for gasoline? Explain.

Cutting price in the hope of increasing demand is one way to boost revenues.
　But in today's glutted market, a reduction in the price of natural gas sold to the United States would be a direct cut in revenues, analysts and industry spokesmen say.

2. What does this say about the price elasticity of the U.S. demand for natural gas? Explain.

The loss of 17 million transit riders because of the fare hike will cost the TTC an estimated $14.5 million.

3. Is the demand for transit elastic or inelastic? Explain.

Another argument that Canada shares with nations caught in the world energy squeeze is that higher fuel bills lead to efficient conservation. The point has yet to be proved. In the past decade, world prices rose from $1.80 to $38 a barrel and Canadian prices went from $2.76 a barrel to $14.75, yet the world gulps more oil than ever.

4. a) Does the author of this article think the demand for oil is elastic or inelastic? Explain.
　b) What's wrong with basing one's judgment on these numbers?

But since Americans will see the benefit of lower oil prices almost immediately at the gas pumps, the U.S. economy will be getting an injection of disposable income resembling a cut in taxes, Ing said. It might speed up economic growth there and that would be good for almost all Canadian industries.

5. What does this say about the price elasticity of demand for oil? Explain.

The province earned about $1.2 billion in 1986 from natural gas production but since markets were unfettered, gas prices have plummeted by 35 per cent and, according to provincial sources, the Crown revenue has dropped even more.

6. What does this say about the price elasticity of demand for natural gas? Explain.

B. Calculating Elasticity

For those who argue that pump prices don't affect usage, look south again. Last year, gasoline sales rose 2.3 per cent in Canada while in the U.S. the guzzling tapered off. There was a five per cent drop in consumption.	**1.** Assuming gasoline prices in Canada were unchanged last year, but U.S. prices rose 20 per cent, what do you calculate the U.S. elasticity of demand for gasoline to be? Explain your calculation.
Industrial power demand could decrease by one or two per cent in response to a five per cent rate increase, said Acton, and it could go down as much as six per cent in response to a 15 per cent rate hike.	**2.** a) Calculate the lowest and highest assumed elasticity of demand when the rate increases by five per cent. b) What is the highest assumed elasticity of demand if a 15 per cent increase occurs? c) Is the six per cent drop in consumption due to a 15 per cent price increase consistent with the earlier figures? Why or why not? d) If the two and six per cent drops in consumption with their respective increases are correct, what does this imply about the shape of the demand curve (i.e. elastic or inelastic)?
Miss Anders's employer, the KFC subsidiary of Heublein Inc., for example, is tightening work schedules at the stores it owns to help offset the effects of the higher wages. In the case of Miss Anders, she will now be working 18 hours a week instead of 23, and earning about $48 a week after Jan. 1, down from the $53 a week she has been averaging.	**3.** a) Calculate the elasticity of demand for labor at Kentucky Fried Chicken. Is it elastic or inelastic? b) Why was it not necessary to calculate the numerical elasticity if you were only interested in whether it was elastic or inelastic? Explain.
In fact, chocolate bar production has been down much more than that of chewing gum. In the first half, chocolate bar sales dropped 31 per cent, vs. a 24 per cent increase in the same 1974 period. The price of a standard-size 1¼-ounce chocolate bar went from 15¢ to 20¢ about a year ago.	**4.** Calculate the price elasticity of the demand for chocolate bars assuming: a) that without the price change there would have been a 24 per cent increase in sales as opposed to a 31 per cent drop; or, b) that without the price change there would have been no change in sales. (Assume that "chocolate bar sales" means numbers of chocolate bars, not dollars of revenue.)

The 17.5 per cent fare increase that went into effect in November has produced only about a 10 to 11 per cent increase in gross revenues rather than the 17.5 per cent that the Taxi and Limousine Commission had predicted.

5. a) Calculate the price elasticity of the demand for taxi service from the figures given in this clipping.
   b) What elasticity did the Taxi and Limousine Commission believe characterized this industry? Explain your reasoning.
   c) What minimum price increase would be required to increase revenues by the 17.5 per cent that was predicted? Explain your calculations.

It estimated wheat use at 2.56 billion bushels, only five per cent less than the record 2.69 billion consumed last marketing year, which will mean higher wheat prices for farmers. The season-average prices are expected to go up between 40 per cent and 50 per cent, between $3.55 and $3.85 U.S. a bushel.

6. Calculate the price elasticity of demand for wheat.

Theatre admissions totalled 1.085 billion in 1988, Valenti said, the 12th straight year attendance topped the one billion mark. The attendance figure was down 0.03 per cent from 1987, Valenti said, while revenues rose five per cent.

7. Calculate the price elasticity of demand for theatre admissions, assuming that in the absence of a price change demand would have been up two per cent due to population growth.

C. Miscellanea

TTC general manager Michael Warren said "there is a psychological resistance to using transit after a fare increase, but people forget about this over time. If the economy improves and there's more money around, then we'll recover our riders quickly."

1. Does this clipping imply that an improving economy will change the elasticity? If so, in what direction will it change, and why? If not, why not?

There has already been enough resistance to bring coffee prices back down from a short runup after the Guatemalan earthquake. At that, futures prices set highs up to $1.04 a pound. The present high is about 98 cents, but it had to come down to about 95 cents in reaction to the sharp runup.

2. Would you say the market initially overestimated or underestimated the extent to which consumers' demand for coffee changes as the price of coffee changes (the elasticity)? Explain your answer.

In anticipation of surpluses the price fell five per cent but sales have increased only one per cent as consumers expect the price to fall even further.	**3.** a) Given the change in sales and price, what is the elasticity of demand? b) What appears to be happening that would lead you to believe that your estimate in a) is wrong? Should it be smaller or larger? Explain.
The prospect is that if the consumer end of the pipeline is unable to halt the price advances, ground coffee will reach $2 a pound. That point, or even one well below it, will mean sharply reduced coffee consumption which, in turn, will mean serious disruption of the entire coffee trade.	**4.** Interpret the statement "if the consumer end of the pipeline is unable to halt the price advances" in terms of the characteristics of the supply-and/or-demand curves.
They note that sales fall off sharply after a major tax increase, and then ease up again as smokers adapt to the new price.	**5.** What does this say about the short-run versus long-run price elasticity of demand for cigarettes?
"Cartels sow the seeds of their own destruction by driving prices to levels that cause new increases of output from new sources."	**6.** What does this say about the long-run elasticity of supply?

I-3 Costs and Perfect Competition

The cost and supply dimension of economic problems in newspaper articles can be analyzed by applying a few basic principles. The points below summarize these principles.

1. The marginal cost (MC) curve passes through the bottom points of the average variable cost (AVC) curve and the average total cost (ATC) curve. This is true because of the arithmetical relationship between marginal and average costs.
2. These costs include opportunity costs, in particular the return that the firm could have made on its assets had they been employed elsewhere. This implies, for example, that the ATC curve includes a "normal return" or "normal profit" which must be earned to keep the firm in this business.
3. In the short run, a firm will continue to operate as long as average variable costs are covered. In the long run, it will operate only if average total costs are covered.
4. The long-run ATC curve is the envelope of the ATC curves associated with plants of all possible sizes.
5. Firms in competitive markets maximize profits by producing the output at which marginal cost equals the market-determined price. Thus the portion of the firm's MC curve above its AVC curve gives that firm's short-run supply curve. Summing these curves horizontally over all firms gives the industry short-run supply curve.
6. In the long run, supply is determined by allowing the size and number of firms to change, and factor prices to adjust.
7. Subsidies and taxes levied on producers can be analyzed by noting their impact on the firm's cost curves.

1-3A Ranching

High rates, low returns hit ranchers

By JOANNE MacDONALD
Sun Business Reporter

High interest loans taken out in the past four years combined with fluctuating product prices have resulted in financial crises for some B.C. farmers, with some being forced to seek jobs outside the farming sector.

Particularly hard hit have been Interior ranchers, many of whom have been forced to liquidate their herds in efforts to keep afloat financially.

Bill Sedgewick considers himself one of the luckier ranchers. But he says his family-run, 640-hectare, 200-head ranch in Vavenby, north of Kamloops, which he operates with his brother and partner, John, is short-staffed. To make ends meet, John has been forced to leave the farm to go logging.

"We're lucky because we can fall back on the forestry industry. But as it is, we have a two-and-a-half person operation that we're running with one and a half people," Bill says.

"To put it quite bluntly, we've had to cut back on a lot of things. There's no extra money. If you've been in business for several years, you probably did improvements, investing capital at 10 or 12 per cent. When interest rates started to hit 20 or 22 per cent, people were shocked. They never expected anything like this."

Grant Huffman, president of the Kamloops-based B.C. Cattlemen's Association, agrees.

"There's a lot of hardship. Female cow numbers (needed to produce new herds) went down drastically in the past four years. But people are still selling down their cow herds and they're not increasing their beef stocks, yet. On the Prairies, farmers can shift commodities from crops to cattle, but B.C. is basically a one-commodity province. The type of land used by ranchers doesn't lend itself to crop changeover," Huffman says.

Lorne Leach, BCCA secretary-manager, has a simple explanation for the current problems facing ranchers.

"What's happening now is all part of what I call the four evils of ranching: high interest rates, low beef prices, high production costs and declining land values. They've led to an over-all disaster for ranchers," Leach says.

---

1. How do high interest rates affect the cost curves of farmers? Why don't the farmers shut down if the high rates and low prices are leading to financial crisis?

2. Why are farmers being "forced to liquidate their herds"? Explain by referring to the cost and revenue curves of a competitive firm.

3. By referring to opportunity costs explain why John is not really forced to take another job but is making an economically rational move?

4. How will the reduction in female cows affect the market place for beef? (sixth paragraph)

5. How does being "basically a one-commodity province" in terms of land use affect the cost curves and thus the supply curve of B.C. farmers when times are bad for ranchers? Could this be one of the reasons why more farmers have not quit raising beef?

6. Of the four evils mentioned in the last paragraph, which is a consequence of the other three? Explain.

I-3B Forests

MB's Smith urges rethink of new forestry charges

The top man at the top B.C. forest products company today urged the provincial government to rethink historic changes imposed last fall in the pricing and management of the province's timber.

"Few could argue against government increasing the economic rent for its forest resource," said Ray Smith, the president of Macmillan Bloedel Limited.

"And there are few companies in the industry that cannot afford to pay the new rates in this phase of the market cycle, although I am told that even in this peak phase there are some operations which find them unaffordable."

The provincial government in September announced new stumpage or timber charges and switched to industry some forestry management costs that until then had been government's responsibility. MB has said that these moves will increase its costs by about $40 million annually.

Smith told the annual meeting of shareholders that unless the government changes the policies "to make them responsive to market forces . . . there will be many more mill closures and more community stress than under the previous stumpage regime." (When the government introduced the new stumpage charges and forestry responsibilities, it said that stumpage would no longer go up and down with market cycles.)

"I urge the provincial government to make sure that it provides a response to market forces comparable at least to that of the previous market-based regime."

---

1. What is economic rent? (second paragraph) What must be the current state of most firms in the forest industry to justify "increasing the economic rent"?

2. How will an increase in stumpage or timber charges affect the cost curves of forest companies? Note: Stumpage fees are taxes levied by the government, based on the volume of timber cut.

3. Explain why some operations would find the higher stumpage fees unaffordable. Hint: To what part of the supply curve would these firms correspond?

4. Using a diagram showing cost and revenue curves, explain why the new stumpage system will lead to "many more mill closures and more community stress than under the previous regime."

Short Clippings

Note: Suggested answers for the odd-numbered questions below appear in the Appendix.

A. Interpreting Cost Curves

He maintains there are three principal reasons why Ontario chicken producers can't compete with an influx of U.S. product. Canadian operations have to heat their facilities whereas most U.S. poultry operations are in the warmer climes of the "chicken belt." Unlike most of their U.S. competitors, Ontario producers also have to deal with unions.

But the chief drawback, says Kurdian, is that U.S. operations are integrated and large. "In the U.S., one operation produces and processes 13.5 million broilers a week. The largest operation in Canada can do half a million a week."

1. a) Explain how each of these three arguments affect
   i) marginal cost;
   ii) variable cost;
   iii) fixed cost, and
   iv) average cost.
   b) What terminology do economists use to refer to the last argument?

Growers, he said, will lose six cents a pound on the 1976 cherry crop. While they cost 37.5 cents a pound to produce, the return from the marketplace and government subsidies will bring in only 31.5 cents.

It is estimated that federal and provincial government subsidies for the 1976 crop will be 13 cents a pound for cherries, 7.7 cents for apricots, 6.5 cents for pears and 6 cents on plums.

While calling the support levels fair, Bernhardt said he was disappointed that limits will be put on the amount of a grower's crop that will qualify for a subsidy.

He said a grower can get a subsidy on only 25 tons of his cherry crop, but one of his members harvests up to 200 tons.

2. a) If cherry producers expect to lose 6 cents per pound, why do they bother producing at all?
   b) If an unlimited subsidy of 13 cents per pound is given to all cherry producers, what happens to the MC, AVC, and ATC curves?
   c) What happens to the MC, AVC, and ATC curves if only the first 25 tons are eligible for the subsidy?
   d) Suppose all cherry producers currently each produce at least 25 tons.
      i) What happens to the supply curve if the subsidy is unlimited?
      ii) What happens to the supply curve if the subsidy is limited to the first 25 tons?
   e) Besides the obvious fact that the limited subsidy will cost less money, does your answer to the preceding question suggest any additional reasons why the government might prefer a limited subsidy?

There are plenty of tanker bargains around. Arne Naess & Co. Inc., a New York ship broker, says prices today are about 10 per cent of what they were at the peak in June 1973. Brokers report the Liberian-registered Benjamin Coates, a 50,000-ton tanker built in 1960, was recently sold for $1.1 million, only slightly above what its scrap value would be.

3. Show explicitly how the existence of a positive scrap value for tankers affects the supply curve for tanker services.

"One reason why government revenue collapsed is that decontrol unleashed a fierce inter-corporate competition for market share," notes industry analyst Wilf Gobert, with Peters & Co. Ltd. of Calgary. "As with any competition, it has led to low prices. Production is only slightly economic in some cases."

4. How would you define, in terms of a cost-revenue diagram, what is meant by production being only slightly economic?

Lang also reported that lumber prices in the United States are now below year-earlier levels. In the meantime, the Canadian dollar has risen against the American, adding to the squeeze on lumber operations.

5. Explain how a rise in the Canadian dollar would lead to a squeeze on Canadian lumber operations.

Young says Pelican's modern machinery, low cost resource, and low labor component (the plant employs 140, about half what an equivalent plywood mill would employ) give OSB a big edge competing with plywood. (Note: OSB stands for Oriented Strand Board.)

6. a) Draw and compare the cost and profit positions of OSB compared to plywood.
b) Explain what will happen in the long-run markets in these two industries. Show how the long-run adjustments will impact on the cost curves drawn in part a).

When the price of feed is low, meat production costs are cut and heavier animals are marketed.

7. Explain how lower feed costs change the cost curves of meat producers so that "heavier animals are marketed"?

"I think there will be continued economic growth, and therefore marginal revenues," Holmes said Monday. "On the employment scene it looks like a gradual continued improvement."

8. By referring to the revenue and cost curves of the firm, explain how increases in marginal revenue lead to more employment.

Wages will be subsidized up to 50 per cent to a maximum of $3.50 an hour, said Richmond. Both private and public-sector employers are eligible.

9. By referring to the cost and revenue curves of the firm, explain how this subsidy leads to the employment of more people.

"This business cycle is the first in the postwar period where profit improvement has not been achieved through price increases."	10. What must have occurred to increase profits?
Last spring, Western farmers planted a record area of land to crops in the face of a worsening economic situation. Analysts said farmers were attempting to compensate for falling prices with increased production.	11. a) In what direction does this analysis have the supply curve sloping? b) Explain why this analysis does not make sense. Hint: Assume that farmers were maximizing profit at the earlier, higher price.

B. Economies of Scale

The sharp drop in the cost of columns printed by daily newspapers as circulation rises does much to explain the relentless tendency in this or any country towards the one-newspaper town.	1. a) What does the cost curve for a newspaper look like? b) Explain how this phenomenon would lead to one-newspaper towns. c) What economic terminology is used to describe this situation?
Davidson said one of the major innovations in Hydro's early development was the rate-structure design, which was promotional in nature. The value of economies of scale was discovered and this later led to quantity discounts for electricity. Hydro soon discovered that the cheaper the cost of electricity, the more demand grew. In the early decades of the century, this helped lead to the economic expansion of the province. But today, he said, the sustained high growth of Ontario Hydro is giving rise to alarm, not only to environmentalists, but to Hydro management.	2. a) What is meant by a "promotional" rate structure? b) How would the "value of economies of scale" be shown on a cost-curve diagram? c) What do you think is causing alarm to Hydro management? How is this shown on your cost diagram?
"Yes, the costs are high, but all parties recognize the difference in economics between a large passenger ship of 1,100 people going to Alaska and a small ship of 80 passengers," he added. "It's in everybody's best interests to work this thing out."	3. a) What is indicated about the long-run cost curve for passenger ships? b) Why would small passenger ships (80 passengers) operate if large ships were more economical?

And they also say that the growth of promising but costly technologies like supercomputers and aircraft have begun to test the classical view of competition among nations because no more than a few nations might ever establish footholds in the industries born of the technologies.	4. Interpret this clipping in terms of cost and revenue curves.
From a production standpoint, the U.S. wine industry could easily drown its Canadian competitors. Newman D. Smith, executive vice-president of Winona, Ont.-based Andres Wine Inc. says: "All they have to do is increase their bottling by six and one-half days. They'll wipe us out."	5. What does this imply about where the Canadian wine industry is on its long-run cost curve compared to the U.S.?

C. Market Reactions

Lee said cutting Canadian power exports entirely would increase U.S. coal production by only one per cent and create just 2,500 more mining jobs. But doing so would also cost U.S. consumers between $283 million and $400 million US a year in higher electricity bills.	1. Show on a supply-and-demand diagram what is going on here.
Prices and profits are good for us now but we have to be concerned about the newly emerging companies.	2. a) Draw the cost and revenue curves facing this firm. (Assume the industry is a price-taker.) b) How will the newly emerging companies affect the market and subsequently this firm in the long run?
In the most extreme case, if every electrical customer in Canada cut power consumption by 10 per cent, the utilities would probably have to raise their rates by close to the same amount to compensate for lost revenue.	3. a) How would the 10 per cent cut be reflected on a supply-and-demand diagram? b) What is implicitly assumed here about the utilities' marginal costs and the consumers' demand elasticity?

The Canadian Press on Monday quoted the chief executive officer of Rolland Inc., another Quebec paper producer, as saying he expects to have to modernize and lay off at least 10 per cent of his 1,400 employees as a result of the competition.	4. a) What appears to have been happening in the paper industry? Show this on a supply-and-demand diagram. b) By referring to the cost and revenue curves of Rolland Inc., show why they are laying off employees. c) How will modernizing make him more competitive?
"Economic logic says this will change," says Maurice. "Markets that aren't very profitable don't necessarily serve the consumer."	5. Using economic terminology explain what Maurice is saying.
Aluminum prices have come down recently to about 75 cents a pound, after rising as high as 96 cents in late summer. Analysts attribute the cool-out to the fact that some marginal aluminum production is getting under way again in response to the higher prices. The metal is expected to settle around 80 cents over the next few months.	6. a) What is meant here by "marginal aluminium production"? b) Use a supply-and-demand diagram to explain this clipping.
A steel company that believes U.S. competitors will be able to treat Canada as a marginal-cost market has launched a drive to improve quality and efficiency, Barrett said, and is modernizing equipment and machinery.	7. What is meant by treating Canada as a marginal cost market?
Lang said he expects lumber prices to remain relatively stable or perhaps improve slightly because current levels are creating difficulties for some Canadian producers.	8. Explain the logic behind this statement. Does it reflect a movement along or a shift in the market supply curve?
If prices decline much, some mills will be forced to close, creating shortages in the market, Lang told Canadian Dow Jones yesterday.	9. Explain why this statement doesn't make much sense.

I-3 *Cost and Perfect Competition* 35

D. Opportunity Costs

Galbraith said the company was not prepared to indicate the location or number of other restaurants it might close as the strike continues. "We are looking at all sites that are not now operating and their potential alternate uses," said Galbraith.	1. What important economic concept is being employed here?
Egg-type chicks, especially bred for laying efficiency, begin to produce at about 20 weeks of age and attain maximum laying rates within a year. Since the cost of raising the chick to laying age represents a large cost to the producer, he attempts to keep the bird laying at a maximum rate as long as possible. Therefore there is a reluctance to retire flocks to the stewing cauldron, unless prices dip so low that continued production is a losing proposition.	2. Explain this argument in terms of profit-maximizing, variable costs, fixed costs, and opportunity costs.
Even though my company is debt free and can afford to pay cash for new equipment, we wouldn't consider it given the high interest rates unless consumers in the market view our industry's product more favorably.	3. a) What does the interest rate have to do with the decision not to buy new equipment even though they can afford to pay cash? b) How would a more favorable view by consumers change his mind?
In recent years, high grain and low meat prices caused many farmers to market grain directly. It was more profitable to sell grain rather than feed it and sell the meat.	4. Explain what is going on in this clipping by referring to the cost and revenue curves of farmers for grain and meat.
"In what is a cyclical and high risk industry, a return on investment of this magnitude is insufficient to provide capital investment in new technology and equipment to remain competitive in world markets," said Ennis.	5. a) How does higher risk affect the cost curves of a firm? b) How will market forces rectify this situation?
The volatility of the forest industry means the average rate of profit of forest companies must exceed that of more stable areas of investment in order to attract investors, he said.	6. How does this show up on a cost-revenue curve diagram?

E. Shutting Down

Prices are an important factor. "In one section of our operations we're within $10 (in the price per thousand board feet) of being better off shutting down."	1. At what point is a firm better off shutting down?
The rates he can command for his equipment barely cover his costs. He needs $80 an hour, for example, to operate profitably the two-year-old Kenworth he drives; he receives, on average, $67 to $68.	2. If he needs $80 and only gets $67 to $68, why does he continue to operate?
The president of one Edmonton-based oil and gas operation confirms that "some of these guys are selling gas just to pay their salaries and to pay the rent. For some it costs more to produce the gas than they get to sell it."	3. Explain why "these guys" are continuing to produce gas if they are not covering their costs.
Owing to the nature of personnel relations in the large Japanese company, with employees essentially hired for their entire careers, all labor costs as well as sales, overhead and interest costs, are, in fact, fixed for the Japanese company.	4. a) Explain why these features of Japanese firms imply that in hard times they will invariably cut prices to compete, in contrast to North American firms who are reluctant to cut prices. Illustrate your answer by drawing a cost-curve diagram comparing Japanese and North American firms. b) Does the argument presented in part a) hold true in the long run as well as the short run? Why or why not?
Sawmills have high debt loads. They need high cash flows to pay the interest on those debts. They may only break even on actual production, but shutting down would deprive them of any cash flow at all. When prices are low, the mills try to reduce unit costs by producing more. It makes sense for an individual mill, but when they all do the same thing, the net effect is to flood the market and drive prices even lower. No one wants to be the first to shut down, because that simply gives the competitors an edge. Each mill operator hopes the guy across the street closes first.	5. a) By referring to the short-run cost curves of a sawmill, explain why they do not shut down. (first paragraph) b) Where would a firm have to be on its marginal cost curve for it to make sense to produce more? c) Explain the last paragraph in terms of the supply-and-demand diagram facing an individual mill.

It also found that almost one-half of the contractors surveyed would leave the business if they could get a decent price for their equipment.

"The reason they feel the way they do is that the rates they get for working their equipment just isn't sufficient to pay the bills, pay proper wages and have a good standard of living," Couiyk said.

6. a) Why can't they get a decent price for their equipment?
   b) If they aren't getting a sufficient rate, why do they continue?
   c) What role does opportunity cost play here?

"The margin which [the entrepreneur] requires as his necessary incentive to produce may be a very small proportion of the total value of the product. But take this away from him and the whole process stops."—**John Maynard Keynes**

7. a) What is meant by the word "margin" in this quote?
   b) What is the "whole process" referred to, and why might it stop if the entrepreneurs' margin is taken away? Explain in terms of opportunity cost.

I-4 The Corn-Hog Cycle

Economic analysis with supply-and-demand curves can do more than just determine equilibrium price and quantity. It can also aid in determining the nature of the sequential process by which a market moves from an old equilibrium to a new one. Such analyses are called "dynamic." The nature of the dynamic adjustment of price and quantity to a market disequilibrium varies widely, depending on the particular market in question. One very important determining factor is the time lag associated with changing supply. In the agricultural sector this is a critical problem, since supply is usually determined well in advance of its delivery, and cannot be changed until the following year's crop is planted. A second important determining factor is the number of suppliers. When there is a large number of suppliers, as is typically the case in agricultural markets, a single supplier is not able to gauge the market's reaction. The result is an uncoordinated supply change which inevitably leads to an overreaction on the part of the market as a whole, creating an undesirable cyclical behavior of prices and quantities.

This illustrates one of the shortcomings of the price system: In some markets random shocks lead to cycles which can sometimes be quite severe, and which always cause headaches and heartaches for those trying to make their living in these markets.* In economic theory, such cyclical fluctuations are referred to by the expression "corn-hog cycle,"** a term that originates from the example first used to illustrate this phenomenon. The term "cobweb" is also used to describe this phenomenon, since tracing dynamic movements on a supply-and-demand diagram usually creates a cobweb pattern, as is illustrated in the example below.

Suppose, in Figure I-4.1, that the D curve represents the demand for pork, the S curve represents the supply of pork and the market is at point A,

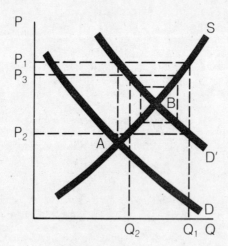

Figure 1-4.1

---

* The government has responded to this problem with policies designed to prevent or cushion these cycles. One of the policies, the development of marketing boards, is discussed in the next section (I-5).

** Because corn is used as feed for hogs, activity in the corn market tended to accentuate the cycle, as described for the hog market.

their intersection. Now suppose, as an example, that there is a random shock to the demand curve, shifting it to D'. The new equilibrium is evidently at point B, but this will not be attained immediately because it takes about a year to raise more pigs to change the supply of pork. Meanwhile, the current supply of pork will be bid up in price to P_1. The price P_1, however, will induce farmers to supply, next year, Q_1 quantity of pork. When this pork reaches the market (next year), there will be an excess supply at price P_1, so the price will be pushed down to P_2. This induces farmers to cut back on pork output, in the following year, to Q_2. But this will cause the price next year to rise to P_3, and so on, creating the cobweb picture in Figure I-4.1 and explaining the cyclical path for pork prices and output.

I-4A Newsprint

Newsprint boom could bring bust

By John Schreiner
Financial Post

VANCOUVER—The surge of newsprint production capacity around the world is casting a threatening cloud over Canada's booming forestry industry.

Encouraged by strong demand, rising prices and burgeoning profits, the world's newsprint companies have embarked on a $10-billion capital spending spree. It's "the largest capacity increase in history," Vancouver industry analyst Charles Widman says. Canada accounts for about a quarter of the new investment.

But the expansion may be laying the groundwork for disastrous overproduction and much lower newsprint prices in the early 1990s.

Vancouver analyst Ross Hay-Roe, editor of the PaperTree Letter, warns that the industry is basing its five-year projections of demand growth on the past five years. And like many economists, Hay-Roe predicts the U.S., market for 45 per cent of the world's newsprint, will slip into recession before 1993.

For the time being, the good times continue to roll for producers. Newsprint prices are at their highest level in real terms since 1921. Canadian producers, who turn out almost a third of the world's newsprint, are hoping for at least one more substantial price increase this year. Some believe there's room for a second in 1989, just before the new mills start running.

"All of our mills are running at full capacity, and we expect this to continue through 1988," Abitibi-Price Inc. Chairman Bernd Koken told his company's annual meeting last month.

The Toronto-based firm is the world's largest newsprint maker. Koken predicted newsprint markets will remain strong throughout this year and into 1989. But he cautioned that supplies will exceed demand by 1991, with softer prices likely for the next two to three years.

World demand for newsprint stands at about 30 million metric tons a year. Almost four more million metric tons will be on the market by 1992, when most of the new machines are running. Yet consumption is expected to grow only half as much as capacity during the next three years.

"The industry is once again in the historical pattern of adding too much capacity at once," MacMillan Bloedel Ltd. pulp and paper Vice-President Eric Lauritzen told a conference of the Canadian and Japanese newsprint associations last month.

Note: Suggested answers to the questions below appear in the Appendix.

1. Draw a supply-and-demand diagram to picture the current state of the newsprint market. Has the current high demand come about by a movement along or a shift in the demand curve?
2. How is the "$10 billion capital spending spree" represented on the diagram?
3. Use your diagram to explain the rationale of the third paragraph.
4. Show on your diagram how the scenario of the third paragraph would be avoided if the projections of the first sentence of the fourth paragraph come true.
5. What does "in real terms" mean? (second sentence in the fifth paragraph)
6. What is meant by "softer" prices, in the seventh paragraph, and why will they come about?
7. By what terminology do economists refer to the historical pattern noted in the last paragraph?

I-4B **Beef**

Beef cycle ties up cattlemen's hands

By Catherine Harris

FARMERS AND consumers who had to contend with the severe beef cycle affecting prices and production in the 1970s hope it won't be repeated in this decade.

Beef production is highly cyclical in that cattlemen increase production when prices are high and liquidate when prices are low. And because it takes three years from the time of breeding to get the animal to slaughter weight, very large supply buildups are possible.

This happened in 1974-77. Supplies were so great that prices plummeted, causing farmers an estimated $400 million in losses, and producing an enormous cutback in production. This, in turn, led to a severe shortage, which caused retail prices to rise 46 per cent in 1978 and 32 per cent in 1979.

The 1970s cycle was more severe than those of the 1950s and 1960s—and there's a lack of agreement about why it happened. Charles Gracey of the Canadian Cattlemen's Association blames government incentive programs in the late 1960s and early 1970s. But Agriculture Canada's Gordon Pugh says it was the grain price explosion which shot feed costs up so much that cattlemen depleted their herds. Both Gracey and Pugh are hoping—and indeed predicting—there will be a gentler cycle in the 1980s.

At the moment, cattlemen are very cautious about increasing production—partly because they were so badly burned last time round, and partly because of rising expenses brought on by high interest rates and growing feed costs.

But this restraint could end if interest rates or feed costs were to fall or cattle prices take off again. And Pugh is worried about a possible price explosion next year because of falling pork supplies and only a very small increase in beef supplies.

1. Explain the third paragraph by using a supply-and-demand diagram.
2. What characteristic of the beef market would make its cycle more severe than that of the pork market?
3. In the fourth paragraph Gracey and Pugh suggest different reasons for how the cycle got started.
   a) By using a supply-and-demand diagram, explain the logic of each of these theories.
   b) If you had data on the behavior of beef prices at the beginning of this cycle, explain how you would determine which of these two theories is more acceptable.
4. Why would beef prices explode if *pork* supplies were to fall, as maintained in the last paragraph? How would this be captured on a supply-and-demand diagram?
5. Would a higher price elasticity of demand for beef intensify or alleviate this cycle? Explain by means of a diagram.
6. Would a higher price elasticity of supply of beef intensify or alleviate this cycle? Explain.

Short Clippings

Note: Suggested answers for the odd-numbered questions below appear in the Appendix.

Hog producers are in the expansion phase of the hog cycle and sharply increased supplies will be available during the coming year.

1. a) What causes hog producers to be in an expansion phase?
   b) What will happen to prices in the coming year?
   c) What will happen in the following year?

A major reason for the expansion has been that market prices of hogs have been high in relation to feed costs for some months. With last fall's record corn harvest and prospects of better profits, farmers plan to feed more to hogs.

2. a) What do you anticipate will happen to the price of pork next year? Explain.
   b) What do you anticipate will happen to the price of corn next year? Explain.

The beef shortage—and the resulting high prices—are expected to remain for several years while Canadian ranchers increase their production, the meat packers said.

3. a) Using the theory of the corn-hog cycle, show on a supply-and-demand diagram where the Canadian beef industry was at the time this item was written.
   b) Why would the high prices be expected to remain for several years? Doesn't the corn-hog cycle theory suggest that prices will fall next year?
   c) Explain the meaning of the term "shortage."

One reason for the higher prices is—ironically—the lower ones.

Kaay said that, because of the temporary price reductions, people are ordering more beef than they normally would, resulting in a shortage which in turn raises prices.

"The amount of available live cattle isn't enough to handle the sudden high demand." He said there has been some "panic buying" by people wanting to stock up on beef while the prices are still low.

4. a) Explain the activity described in terms of a supply-and-demand diagram.
b) How does this situation relate to the corn-hog cycle?

With the current price drop due to increased supplies we will see more customers but some will hold off buying in the hope prices will fall further. Those who hold off will probably end up paying more when they come into the market to buy.

5. Explain what is happening and why those holding off will pay more. Use a supply-and-demand diagram in your answer.

Beef prices are rising and are expected to stay high because of an increased demand for Alberta beef in the United States, a B.C. Cattlemen's Association spokesman says.

"There is a good demand from the U.S.," said Henry Blazowski, the association's secretary manager. "Meanwhile, the supply isn't that great because we had to pay our bills with something when the prices dropped and so we slaughtered our herds."

6. What economic theory would best describe this scenario? Explain.

Lower grain prices are encouraging increased production of all meats. The year ahead looks promising for consumers but very dismal for producers.

7. Explain the chain of events in this clipping that leads to meat producers hurting themselves.

An industry official says Canadian newsprint mills have created both a major threat and a challenge to themselves by "over-reacting" to high demand.

8. What is this over-reaction, and why is it a threat?

"Obviously, we have not learned our history lessons in the paper business. I consider this onslaught of new capacity to be not only the biggest threat but also the biggest challenge to the Canadian industry," he said.

9. a) What is the history lesson referred to?
b) Why is the new capacity a threat?

---

44 Part I *Microeconomics*

"It is difficult to understand why, when business is good for a short period, many Canadian newsprint producers jump aboard the increase capacity bandwagon with the inevitable result that over-expansion produces surpluses in the market place."

Wiewel said if historical patterns prevail, the surplus will have its main impact on Canadian operating rates.

10. a) What is meant here by surpluses?
b) The clipping refers to Canadian operating rates. What other impacts are there? How is this shown on a supply-and-demand diagram?

Car makers are already experiencing a cyclical decline in demand which is normal after several years of growth. It will lead to a short-term weakening in the auto sector, said Rheaume.

But new production capacity, mainly from foreign auto companies opening plants in North America, will trigger a shakeout in the industry in 1990, said Rheaume. Plant closures are expected.

11. Explain this clipping by showing what is happening on a supply-and-demand diagram for cars.

Commodity producers generally, whether they produce wheat in Saskatchewan, copper in British Columbia or natural rubber in Malaysia, have been through five lean years. We, the consumers, were kings of the castle.

Now it's time for the laws of supply and demand to give us the inevitable kick in the pants.

The higher-cost and less-efficient producers having been driven out of business, the first half-decent economic pickup that comes along is enough to run available supplies short.

12. Use a supply-and-demand diagram to explain this clipping.

But in the midst of the boom, some executives worry about a possible glut. Healthy demand and high capacity levels have producers talking expansion again, and some fear the forest industry may be about to spoil its own party by bringing too much new capacity onstream. To John H. Carroll, a vice-president of Andras Capital Research Inc. in Montreal, the industry appears "well on the way to shooting itself in the foot again, as companies join the expansion bandwagon." A newsprint glut would have serious ramifications for earnings.

13. Use the theory of the corn-hog cycle to explain this clipping.

I-5 Marketing Boards

In the previous section we discussed the corn-hog cycle, a market phenomenon characterized by large swings in prices and output, and caused by producers, particularly in the agricultural sector of the economy, determining the next period's output on the basis of this period's price. This results in gluts and shortages, and creates market instability which creates hardships for both producers and consumers. When a glut develops, prices drop and producers go bankrupt or suffer a severe loss of income; when shortages develop the price rises and consumers complain of being ripped off.

Although most economists favor free competition because of its efficiency, they usually admit that unstable competitive markets are undesirable to the degree that government intervention in or control over such markets to prevent instability is needed. The loss in efficiency caused by the government intervention is more than offset by gains to society from having a steady supply of food (for example) at stable prices. Agricultural markets are renowned for their instability, for reasons discussed in Section I-4. Farmers tried to deal with this problem by forming cooperatives, but the voluntary nature of these organizations caused them to fail. The government has now set up marketing boards to deal with this problem.

Just because many economists admit that government intervention in or control over unstable markets is needed, however, does not necessarily mean that they support the methods actually used by the marketing boards to deal with this instability problem. In fact, many do not. Their basic complaint is that many marketing boards do not employ methods of stabilization that minimize the efficiency loss entailed by restricting the operation of free competition. Although this could result from stupidity or poor forecasting on the part of the marketing boards, it usually results from the adoption by the marketing boards of extra goals, beyond the stabilization goal. One popular goal is redistribution of income from consumers to farmers; another is maintenance of the "way of life" of small farmers. Attempts to meet these extra goals involve further distortion of the competitive market, further reducing economic efficiency.

Marketing boards do not all operate in the same way, since they must tailor their operations to the particular market with which they are dealing (the wheat market is more strongly influenced by international forces than is the egg market, for example), but they can in general be characterized as monopolies with the power to control supply and in some cases prices. Supply is controlled by preventing free entry to the industry through the use of output quotas for producers. (These quotas are usually determined, at the time of the marketing board's formation, by farmers' current output levels.) The quotas are valuable, and a market develops for the quotas, with the market-determined price of a quota right depending on the profits that can be made from producing that quota of output. (A quota right is the legal permission to produce a certain level of output.)

These profits in turn depend on the price of the output (often determined by the marketing board) and the efficiency of the producer. If the price is determined by the marketing board it is usually set at a level such that an average-sized farm will make a reasonable return. In a competitive market the smaller, inefficient farms will be forced out of the market and the remaining farms will expand to the optimal farm size; the price will consequently fall, forcing out of business all those farmers who do not move to the optimal farm size. With a marketing board, however, although the small or inefficient farms may be forced out of business, there is little pressure on the less-than-optimal-size farms to increase to the optimal size (independent of whether or not the marketing board sets the price). This is because to increase his output a farmer must first buy a quota right. The money paid for a quota right could be invested elsewhere to earn a return approximately equal to the return from owning the extra quota right and producing the extra output. As a result, there is little incentive for producers to expand to the optimal farm size; thus, the industry is not as efficient as it could be.

Inefficiency usually manifests itself in a higher-than-necessary price to consumers, and sometimes also appears in the form of surpluses of agricultural products. The disparity between the Canadian and U.S. price for eggs, the selling of Canadian powdered milk on world markets at extraordinarily low prices, and the rotting of huge numbers of eggs in storage are instances of the costs associated with the marketing board system. Such spectacular examples inevitably lead to newspaper comment.

I-5A Poultry and Eggs

Marketing boards hit by economists in West

TORONTO (CP)—Canadians are paying roughly $1 billion a year too much for eggs, chicken and turkey because of marketing boards, a study by two Western Canadian economists says.

The study by University of Lethbridge professor George Lermer and W.T. Stanbury of the University of B.C. concluded that dismantling the boards, which control the supply of eggs and poultry through farm quotas, would save consumers between $704 million and $1.04 billion a year.

The economists said the boards should be abolished and the federal government should pay farmers $1 billion for the current value of their production quota, which they received free when the boards were established.

"We are saying the government should pay (them) off and be done with it," Stanbury said, adding that marketing boards have inflated poultry and egg prices to include the cost of production quotas.

Under legislation governing the boards, farmers who want to go into the egg or poultry business must purchase a quota—the right to produce—from an established farmer.

However, David Kirk, executive secretary of the Canadian Federation of Agriculture, said abolishing the boards "would throw the industry into the old cycles of incredible insecurity," adding that producers would lose their protection against foreign competition.

Note: Suggested answers to the questions below appear in the Appendix.

In Figure I-5.1 (a) and (b) suppose that before the marketing board is created the market for poultry or eggs is characterized by the typical farm producing q_0 at a price P_0, with a total supply of Q_0. Now suppose that the marketing board introduces a quota system, permitting the typical farmer to produce only q_1, moving the total output from Q_0 to Q_1.

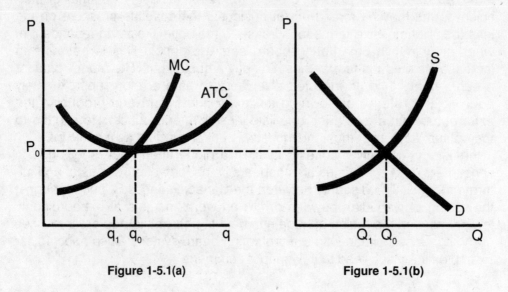

Figure 1-5.1(a)   Figure 1-5.1(b)

1. What happens to the market price due to the creation of a marketing board? Show this on Figure I-5.1(b).

2. What happens to the typical farmer's income? Quantify your answer by using Figure I-5.1(a).

3. Lermer and Stanbury give a rough estimate of the cost to consumers because of marketing boards as opposed to a free market. Show how they would arrive at their total figure by using Figure I-5.1(b).

4. If you have studied consumer and producer surpluses, show what has happened to both, using Figure I-5.1(b), due to the creation of a marketing board. See introduction to I-6.

5. The article indicates (fourth paragraph) that for a new farmer to produce eggs or poultry he or she must buy an existing quota. How would the price of this quota be determined? Hint: see question 2.

6. Why doesn't the marketing board just *give* new farmers quotas?

7. Why would farmers who received their quota free include this in the cost of production?

8. How and why does David Kirk (last paragraph) feel that abolishing the boards "would throw the industry into the old cycles of incredible insecurity"?

It costs $100,000 to milk a dairy farm

By John Spears Toronto Star

So you want to be a dairy farmer. Well, step up to the window and plunk down $100,000.

That's how much it's likely to cost you to get into the game—before you buy a single cow, barn, milking machine or acre of land.

The $100,000 is about what you'll have to pay for "quota"—the right to produce milk in Ontario—if you're just breaking into the business.

In fact, it could well cost you more, and even Ontario's dairy industry is beginning to wonder if that's too much.

Ken McKinnon, chairman of the Ontario Milk Marketing Board, suggested in a speech at the board's annual meeting in January that he'd like to see the cost of milk quotas return to "more reasonable levels"—perhaps 25 per cent lower.

And although there's no evidence to confirm it, one dairy farmer on the meeting floor wondered aloud whether some sharp operators are speculating in quotas—turning a profit by buying low and selling high.

Are high quota values preventing new farmers from getting into the dairy business? And is the high cost pushing up the price of milk for consumers?

The sky-high cost of quotas isn't necessarily the culprit, says Murray MacGregor, an agricultural economist at the University of Guelph.

Instead, the high quota prices may be evidence of inefficiency built into Canada's agricultural industry.

The quota issue stems from Canada's milk marketing system; it sets limits, or quotas, on the amount of milk dairy farmers are permitted to produce. A farmer who produces more than his or her quota is hit with a financial penalty.

Quota is in effect a licence to produce milk, so it has an economic value in much the same way as a taxi licence. And like a taxi licence, it can be bought and sold.

At this year's milk marketing board convention, McKinnon noted that most dairy farmers would like to produce more milk. Strong demand has put pressure on the prices for quota, he said in a speech.

But MacGregor thinks that high quota prices may be the result of the move to fewer dairy farms, rather than the cause.

A lot of middle-sized dairy farmers have been expanding and buying up quota, MacGregor suggests.

Because the amount of quota available to them is limited, the extra demand drives up the price.

MacGregor figures that if some farmers are willing to pay a higher price for quotas, it must be because they're making good profits—probably because they're cutting their costs through good farming practices.

But he doesn't think the high quota price is driving milk prices higher.

If dairying were profitable and there were no quotas in place, the value of the farm assets would be bid sharply higher, so it would cost just as much to break into the business, he argues.

What MacGregor sees as the weakness of the dairy system is that Ontario farmers who want to expand can't buy milk quotas from farmers in other provinces, notably the Prairies, where quota is extremely cheap.

The dairy marketing system gives each province a share of the national market, and few provinces want to give up their market share.

So despite the fact that many Ontario cheese and chocolate makers can't get enough milk to run their plants at capacity, there's no legal way for Ontario farmers to supply them with extra milk.

It would make sense, says MacGregor, to let Ontario dairy farmers buy quota from farmers in other provinces—notably the West, where quota prices are much lower, presumably because dairying there isn't as profitable.

That would let Prairie farmers concentrate in areas best suited to the West, such as grains and beef cattle—and permit eastern farmers to produce more dairy products, where they seem to have a natural advantage because of climate as well as closeness to major markets and processing plants.

Reprinted with permission of The Toronto Star Syndicate.

1. Suggest how McKinnon could make the cost of milk quotas lower. Explain. (fifth paragraph)
2. What must speculators in quotas expect the future policy of the marketing board will be? Explain. (sixth paragraph)
3. Are the high quota values the cause or effect of higher milk prices? Explain. (seventh paragraph)
4. How could quotas lead to inefficiency in Canada's agricultural industry, as claimed in the ninth paragraph? Explain by referring to Figure I-5.1(a).
5. Explain how "Strong demand has put pressure on the prices for quota" as noted in paragraph 12.
6. Explain the logic of the thought expressed in paragraph 13.
7. Evaluate MacGregor's argument as it is expressed in the sixth paragraph from the end.
8. Why do you think that quotas are cheaper in the West? Explain.
9. If Prairie farmers have a natural advantage in grains and cattle, why are they not moving out of the dairy industry?

Short Clippings

Note: Suggested answers for the odd-numbered questions below appear in the Appendix.

A. Interpretation

More farmers and industry officials are seeking cooperation from growers in an effort to get higher prices.	1. What cooperation do you think is sought?
The provincial egg marketing board's uneven distribution of egg production quotas to farmers further contributes to delays in selling eggs, he said. Burrows cited the example of Fraser Valley eggs having to be shipped to Prince George because egg producers in the north do not have enough quotas to fill local demands.	2. If excess demand exists in Prince George why doesn't the marketing board allow producers there to have more quota?
Incentives to produce? Support prices have acted as an incentive to overproduce. Marketing boards try to avoid that problem by aligning supply with demand. The early experience of the Canadian Egg Marketing Association, however, shows that they don't always succeed. Too many yolks is a heavy yoke to bear.	3. a) Why would support prices act as an incentive to overproduce? How would this be illustrated on a supply-and-demand diagram? b) Explain the meaning of the last sentence.
It's a seductive argument. Supply-management schemes and price-support payments foster competition by ensuring that severe price fluctuations don't ruin smaller operators, leaving the field to the big producers.	4. a) Draw a supply-and-demand diagram to illustrate a "supply management" (output quota) policy, and another diagram to illustrate a "price-support" policy. Note for each case the price and output that would result. b) What government policy would have to be implemented in regard to exports or imports to ensure the success of either of these programs? c) Using the results derived in a) and b), comment on the claim in the clipping that competition is fostered by these policies.

"Professional associations will adamantly oppose moves to legalize advertising by their members, but that would seem to be because these associations are administrative arms of cartels, not because the 'public interest' is best served by restrictions on advertising."

5. a) What is a cartel?
   b) In the context of this statement, exactly who are being thought of as cartel members?
   c) Explain in your own words the logic that lies behind this statement.

Marketing boards go back a long way in Canada and are of quite different types.

Some simply promote the product grown by their members. Others try to establish "countervailing power" for farmers who otherwise would each face, on his own, the concentrated power of corporations buying products for resale to consumers.

These countervailing power boards, such as the Canadian Wheat Board, have Economic Council approval.

But the report is heavily disapproving of boards which go one step farther and try to. . . .

6. Complete the last sentence.

"Because," says Timothy, "it was a rat race. Everybody was cutting everybody else's throat. There'd be a big surplus of oysters and the fish wholesalers in Vancouver took advantage of everybody."

7. What part of this clipping is a good argument for an oyster marketing board?

Wise said butter manufacturers will have to become more attuned to market demand and not produce when there is no demand. Otherwise, they'll pay the storage costs.

8. Why would butter manufacturers now produce when there is no demand?

B. Quota Values

We have been through the experience of withdrawing farm subsidies. The clear evidence is that subsidies and tax concessions are capitalized into asset values. When subsidies were withdrawn, farm prices fell 40 per cent.

What is clear from this is that taxpayer-funded subsidies were driving up the cost of basic inputs to farming. They were keeping out new entrants while delivering big capital gains to those who were early entrants to farming.

1. What does "capitalized into asset values" mean? How does it come about?

The amount any farmer can produce is limited by his "quota" issued by the marketing board.

The quotas are issued free to the original producers but later they are bought and sold on the open market since anyone owning the quota is allowed to deliver the stated amount of produce at the protected price.

2. What price for a quota would the open market produce? Explain.

Egg farmers, the report says, are willing to pay this "membership fee" because of the benefits that membership in the cartel confers on them. And the benefits have been increasing sharply—or at least the cost of quota has, according to the study.

3. a) What is the "cartel" referred to here?
   b) What is the "membership fee"?
   c) Explain why increasing benefits would be equated with the cost of quotas as implied in the last sentence.

"In addition," the Veemans noted in a recent presentation to the Agricultural Institute of Canada Conference, "problems arise from the tendency for program benefits to become capitalized into the value of quotas."

4. What does it mean to say that the program benefits "become capitalized into the value of quotas" in the context of marketing boards? What problems arise from this tendency?

According to a report published by the World Bank last month, all developed countries are unnecessarily subsidizing their farmers, either directly by cash or indirectly by quotas and marketing boards, by $46 billion U.S., while all developing countries are doing the same to the extent of $18 billion U.S.

The principal beneficiaries of these vast subsidies are, according to the World Bank:

• Eastern Europe, which imports food, principally from Western Europe, at a saving of $23 billion U.S. to itself (paid by Western taxpayers);

• Landowners and the banks, which have loaned them the money to buy the land, and whose properties increase in value proportionate to the subsidies given to farmers.

5. Why would property values increase in proportion to the subsidies given to farmers? Explain by means of a supply-and-demand diagram.

The European Community, very tentatively and very nervously—because there is nothing governments fear more than long picket lines formed by farmers riding their tractors—has begun to ratchet back its subsidies.

The mountains of beef and grain and butter, and the lakes of wine, will be around for a long time yet. But they may all cease to grow exponentially, as has happened already to the lake of milk (because of lower quotas). Already, land prices in Europe have dropped by an average of 20 per cent.

Some farmers even have complaints—especially those who don't like the high prices they pay for production quotas, or licences, to produce poultry, eggs and milk.

Since output of the various commodities is fixed by the boards, demand for quotas is strong, and prices high: in Ontario, a licence to produce one litre of table milk a day for a year currently costs $275. For the average dairy farm producing 500 litres a day, the cost of a new quota would be $137,500—a prohibitive price for any newcomer wanting to start a decent-sized dairy farm.

The Canadian boards try to set production quotas at a level which ensures that producers receive a return above the cost of production. The Economic Council of Canada calculated in 1981 that quotas added $15,000 annually to each egg producer's income and $23,000 per year to the income of broiler producers.

Harvey said the high price of milk set by the board results in the high quota price, which is passed on to consumers. He said there was no effective consumer lobby because "it's such a complex system, the consumer doesn't really understand it."

6. a) Use a supply-and-demand diagram to explain what caused the mountains and lakes, and why the lake of milk has stopped growing.
   b) Explain why the price of land has fallen.

7. a) Explain why farmers must pay high prices for production quotas.
   b) If the interest rate were 10 per cent, what would you guess would be the extra annual income, due to the quota system, for a farmer producing 500 litres of milk per day?

8. Does the high price of milk cause the high quota price (as implied by the first half of the first sentence) or does the high quota price cause the high price of milk (as implied by the second half of this sentence)? Explain your reasoning.

I-6 Government-Fixed Prices

A common government policy is to fix the price of a good or service. This action runs counter to all economic theory which states that only through freely fluctuating prices can the price system allocate and distribute goods and services efficiently. Sometimes the government fixes the price mainly to stabilize it, as is the case in the example of a fixed exchange rate. But more often the price is fixed because the government feels the market-determined price is either too high or too low.

A very successful investment counsellor once revealed that his major strategy was "betting against the government": He simply found a market in which the government was attempting to fix the price of something (for example, gold) and then invested on the supposition that the government would be unable to maintain that price. This is one major problem with government price-fixing: usually the economic forces are too strong for the government to contain. The second major problem associated with price-fixing stems from economic theory. With a price fixed either too high or too low (relative to the market-clearing price), allocation and distribution of goods and services will be inefficient—the economy will not be as well off as it could be. If the price of a resource such as labor is fixed too high, not enough labor will be employed, resulting in underutilization of available labor. When the price of a good is fixed too high, too many resources are allocated to the production of the good in question, resulting in unwanted surpluses (some of the resources should have been used to produce other goods in stronger demand). When the price is fixed too low, not enough resources are allocated to its production, resulting in shortages (and rationing if imports are not available).

This misallocation of resources is sometimes pictured by using the concepts of consumer's and producer's surpluses. Consumer's surplus is the difference between what a consumer would be willing to pay for the quantity bought and the amount actually paid for that quantity. The amount he would be willing to pay for the quantity bought can be conceptualized by pretending the consumer were charged the highest price he would pay for the first unit, then the highest price he would pay for the second unit, and so on.

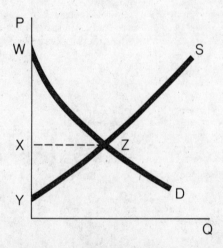

Figure 1-6.1

56   Part I   *Microeconomics*

Graphically this is given by the area under the demand curve to the left of the quantity purchased. Thus consumer's surplus is given by the triangle WXZ in Fig. I-6.1. Producer's surplus is the difference between revenue and costs of production, shown in Fig. I-6.1 by the triangle XYZ. Although the concept of consumer's surplus is not held in high regard by most economists (they don't like anything that pretends to quantify utility), it is often useful in illustrating misallocation of resources and the effects of government policies.

I-6A Minimum Wages

Minimum wage law no help to unskilled

By Walter Block

VANCOUVER

TERRY SEGARTY, British Columbia's minister of labor, has a problem, and is resolved to do something about it in the fall.

British Columbia has the lowest minimum wage rate in the country—$3.65 per hour—and tremendous pressure has been placed upon him to raise it at least to $4 per hour, the average level obtaining in the other nine provinces.

At first blush, this would seem like a good idea, even one that is long overdue. If, as its name implies, the minimum wage law can boost wages up to whatever level is prescribed, that is to say set a floor under incomes for the poor, then why not?

But a moment's reflection will show that this is a mirage. For example, if prohibiting compensation below some arbitrarily determined level can really enhance salaries, why stop at the paltry, mean and niggardly $4 level? Why not go for, say, $40 per hour, or even better yet, really reach for the stars and demand that no employee be paid less than $400 per hour?

The answer is obvious. To mandate that a skilled craftsman with a productivity level of $25 be paid $400 is to invite disaster. Any employer who complied would rack up $375 per hour in red ink. Even at the more modest $40 per hour, any such firm would still lose $15 per hour—and thus be forced into eventual bankruptcy.

No, the reason wages are as high as they are has nothing whatever to do with legal compulsion. It is because productivity is relatively great in this country and because salaries tend to be equal to productivity levels that we enjoy our relative prosperity.

True, a minimum wage level of $4 would not threaten the livelihood of the person who can produce $25-worth of goods and services per hour, but it certainly can put at risk the jobs of people with lesser skills. For example, the employment of a person who can only create goods valued by the market at $3.25 per hour would be obliterated by a minimum wage of $4 per hour.

How can we test the economic principle that high minimum wage levels lead to relatively increased unemployment rates for unskilled workers? One way is to calculate the unemployment rates of youthful Canadians as a percentage of those of the more highly productive adult employees, and compare them with the minimum wage levels in each of the provinces. (For our table, we choose workers between 20 and 24 as our control because this is the youngest group subject to the "adult" minimum wage law.)

The results are painfully obvious. Manitoba, with the highest minimum wage level ($4.30) has an unemployment rate for its young workers that is 1.9 times as high as that for the rest of the population. Saskatchewan, with the next greatest level ($4.25), weighs in with the second biggest relative unemployment rate for youth—1.6 times as high as the rest of the population. And at the bottom of the pack, in terms of the disenfranchisement of their young people, come British Columbia and Alberta with two of the country's lowest minimum wage levels.

Are you listening Mr. Segarty?

	Unemployment rate for 20-24 year olds as % of rate for those 25 and over	Minimum wage
	%	$
Manitoba	289	4.30
Saskatchewan	257	4.25
Ontario	251	4.00
New Brunswick	237	3.80
Nova Scotia	213	4.00
Quebec	206	4.00
Newfoundland	204	4.00
British Columbia	190	3.65
Alberta	182	3.80
Prince Edward Island	n.a.	3.75

Source: Statistics Canada, Labor Department, May, 1985.

WALTER BLOCK is senior economist at the Fraser Institute, Vancouver.

Note: Suggested answers to the questions below appear in the Appendix.

1. If the table in this article had reported the overall provincial unemployment rates in the first column, the results would have been quite different; in particular, the higher minimum wages would have been matched with provinces with lower overall unemployment. In light of this, how would you defend the use of the numbers reported in the first column?

2. Draw a supply-and-demand diagram of the labor market with "number of young people" on the horizontal axis and "wage rate" on the vertical axis. Assume that the labor market is initially in equilibrium and that then the government enacts legislation requiring a minimum wage rate above the market-clearing wage. Draw in this minimum wage on your diagram.

   a) How would you measure the level of unemployment on your diagram? Has it risen or fallen as a result of the minimum wage legislation?
   b) Do the figures tabulated in this article support this result?
   c) By how much does the level of employment change?
   d) The legislation clearly raises the wage rate. Does it also raise total wage income? (Hint: Your answer will depend on a particular elasticity.)

I-6B Rent Controls

Landlords lobbying for shelter allowance to replace rent rules

By **David Israelson** Toronto Star

Ontario's rent controls should be replaced by shelter allowances for low-income tenants, a landlords' lobby group says.

Controls "cater to the well-to-do," while allowances are "flexible and humane," the Fair Rental Policy Organization of Ontario said.

Allowance programs are "run successfully in British Columbia, Manitoba, New Brunswick and Quebec," the group says.

The organization, which represents landlords for about 200,000 rental units across Ontario, released a statement yesterday calling on the provincial government to streamline its $30 million-a-year rent review system.

The statement said:

☐ Rent review should be replaced with a law that prevents "unconscionable" rent hikes, to eliminate the complicated formulas now used to decide on increases.

☐ The streamlined law should be coupled with an allowance program. Under such programs, governments decide how much of a tenant's income should go toward rent and make up the difference when the percentage gets too high—for example, above 30 per cent of income.

House prices up

The demand for reform comes as Metro's tight housing market appears to be getting tighter, with apartment vacancies near zero and another wave of near-panic house buying.

The landlords' group said studies show that shelter allowances would work better than rent controls to achieve the government's aim of helping the needy.

The most recent Canadian study of allowance programs, from 1985, found such programs give tenants more choice on where they want to live.

1. Explain, on a supply-and-demand diagram, why landlords prefer to have an allowance program rather than a rent controls program.

2. How will allowance programs "give tenants more choice on where they want to live" as stated in the last paragraph?

3. Suppose Fig. I-6.2 represents the supply-and-demand situation for rental housing.

   a) Which price-quantity combination best represents the current situation with rent controls?
   b) Is there a shortage or surplus of housing? Identify it on the diagram.
   c) Would you expect discrimination (racial discrimination, for example) to increase or decrease when controls are enacted? Explain why.
   d) In terms of the diagram, how big is the consumer's surplus under controls? How big is the producer's surplus?

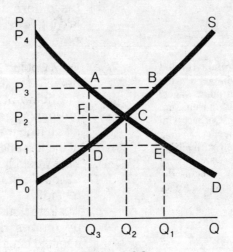

Figure 1-6.2

I-6 *Government Fixed Prices* 59

Short Clippings

Note: Suggested answers for the odd-numbered questions below appear in the Appendix.

A. **Minimum Wages**

At $4 an hour some employees are not productive enough and they are laid off, and employees are selected that hopefully will be able to produce $4 an hour in work. To increase the minimum wage would put many more employees in the unproductive category, and therefore they would be laid off.

1. a) What rule for profit maximization does the first sentence describe?
   b) Use a supply-and-demand diagram to illustrate the impact of a rise in the minimum wage described here, and identify on the diagrams the employees laid off.

But when a government-funded study warns of the dangers in the upward spiral of minimum wages, that's news. Just such an unusual event has happened in Quebec, with the release of a document contending that the province—and particularly its younger workers—are paying heavily for the policy of being the home of the highest minimum wages in North America.

2. a) Why would Quebec be "paying heavily for the policy of being the home of the highest minimum wages"? Doesn't this mean higher incomes for people at the low end of the pay scale?
   b) Why would the younger workers be particularly of note in this regard?
   c) Illustrate your answer to part a) by means of a supply-and-demand diagram.

An impact study of the largest increase in the B.C. minimum wage—a 33 per cent jump from $1.50 to $2 in 1972—showed that of 38,000 employees in 1,029 business establishments, only 364, less than one per cent, were laid off.

3. Explain why these numbers are misleading. (Hint: How many of the 38,000 experienced an increase in wages as a result of the rise in the minimum wage?)

Ron Marcoux, executive vice-president of McDonald's Restaurants, would not comment on how McDonald's would be affected by an increase in the minimum wage without knowing the exact increase.

He said McDonald's pays, on average, "substantially higher" wages than the minimum wage of $3.65.

4. Do you think this means that the typical McDonald's worker is making "substantially" more than the minimum wage? Explain why or why not. (Hint: Consider the arithmetic of computing the average wage, given that no one makes less than the minimum wage.)

Students can expect a harder time finding their first job under B.C.'s new $4.50 minimum wage, an accommodation industry spokesman says.

5. Why would students be singled out here?

"Manitoba has been at $4.70 since 1987 and has retained one of the lowest unemployment rates in the country, so those who try to equate a lower minimum wage with employment are talking hogwash," he said.

6. How would you rebut this argument? Hint: Read I-6A.

B. Rent Controls

"In many cases rent control appears to be the most efficient technique presently known to destroy a city—except for bombing."

1. Explain how rent control can destroy a city.

Rent controls did not dry up the supply of rental accommodation, according to land economist David Baxter.

The factors that created today's rental market are dramatic demographic and social changes, including female equality, the coming of age of the post-war boom babies and the advent of two-income households, he said.

2. a) Explain what impact the factors cited in the second paragraph would have on a supply-and-demand diagram.
b) What does your answer to part a) suggest will happen in the rental accommodation market if:
   i) rent controls were in force?
   ii) there were no rent controls?
c) What does your answer to part b) suggest concerning the validity of the statement in the first paragraph of the article?

If renters were subsidized to the same degree that home owners are subsidized by government, and every renter were given $5,000 to spend on rental accommodation, "you'd have a supply (of rental accommodation)," said Baxter.

3. Explain how government subsidies to renters would bring about a supply of rental accommodation.

"Controls are necessary in some areas of the province where the vacancy rates are not high enough for the market to behave in a normal way. There could be rent gouging and controls would be difficult to repeal so long as the market is in its current condition."

4. What do you think is meant by the market behaving "in a normal way"? How would the imposition of rent controls solve this problem? How would the imposition of rent controls prevent the market from behaving "in a normal way" according to the usual meaning of this term?

Controls were imposed in the "temporary rent regulation measures act" when there was a zero vacancy rate in the province.

"But the government believed a free market would take care of itself and said they'd stay in for 18 months," said Cavanagh.

"But controls didn't stop people from building because anything new was never under control."

5. What is meant here by a free market taking care of itself?

Other things equal, an increase in the interest rate leads to a decline in all long-term capital investments. But why then have interest rates not punched as serious a hole in the construction of office buildings, factories, warehouses and commercial space? While vacancy rates for residential rentals hovered at zero, empty space in office towers amounted to about 8.9 per cent. A similar analysis can be applied to rampaging building costs: they harm all building but commercial non-rent-controlled construction seems, curiously, almost exempt.

6. Is the interest elasticity of rental accommodation supply greater or less than the interest elasticity of commercial space supply? Although there are several reasons for this difference, the author suggests only one. What is this reason?

Immigration and population gains can be the occasion for temporary reductions in vacancy rates, but markets can usually be relied upon to adjust over the long haul.

7. Explain how markets adjust over the long haul to accommodate immigration and population gains. What undesirable short-run effects will occur?

Patterson said that, although Greater Vancouver now has quite a low vacancy rate, 1.6 per cent, there is a high rate in more expensive accommodation and the government could consider starting with control removals near the $500-a-month cut-off point.

Units which rent at more than $500 a month are exempt from the provincial rental increase ceiling of 10.6 per cent per year.

8. Explain why there is a high vacancy rate in the more expensive rental accommodation but not in the cheaper rental accommodation.

In addition, some areas have quite high vacancy rates and could be removed from the controls, Patterson said.

A vacancy rate of three to four per cent in any sector or region would allow consideration of removal of controls, he said.

9.
a) Explain why the rent control commissioner would feel that areas with high vacancy rates could be removed from controls.
b) If you were to draw a supply-and-demand diagram for rental housing with rent controls in force, how would this diagram differ for the case of high vacancy rates versus low vacancy rates? Is this consistent with your answer to part a)?

"Interprovincial migration is expected to continue to exacerbate shortages in affordable housing. Although there are signs of an abatement in the flow of migrants to Ontario, below-normal vacancy rates in many parts of the province will persist. Rising construction costs, combined with rent controls, continue to limit the supply of rental accommodation."

10.
a) Show the impact of interprovincial migration on a supply-and-demand diagram. Hint: Make sure your diagram incorporates rent controls.
b) How would your answer to part a) be changed if rent controls did not exist?
c) How do rising construction costs affect your diagram, and affect the shortage?

C. Miscellanea

"A man went into Barrett's store to buy some potatoes. 'How much is a sack?' he asked. '$4 a sack,' replied Barrett. 'But I can go down to McNamara's and get a sack for $3.50,' said the man. 'Well, why don't you just do that?' asked Barrett. 'Because McNamara's got none left,' said the man. 'Well, when I've got none left, I'll charge $3 per sack,' said Barrett. And that's how it is with oil prices."—**John Crosbie, federal finance minister**

1. Explain what this says about Canada's oil pricing policy.

If banishment of the government's heavy hand is to make us free, we must repudiate state controls however attractive they may superficially appear to be. We must resist the demand of tenants for rent controls, of builders for wage controls, of unions for price (but not wage) controls, of farmers for marketing controls, of manufacturers for import controls.

2. Why would the writer feel that state controls are superficially attractive?

Price controls may work in the short run to prevent price-gouging, but as the population grows in the long run the gap between the market price and controlled price will reach a point at which it will be impossible to continue.	**3.** a) How do price controls prevent price-gouging? b) Why will it be impossible to maintain price controls?
He acknowledged that the effort to scrap all controls "will be a difficult and contentious issue." But, he said, "this is the best time to deal with the issue. We have plenty of gas; we're not in a shortage situation."	**4.** According to the writer what will happen to the price of gas if controls are scrapped? Explain.
Attempts to break the black market's hold on basic commodities have also failed because young people line up all day for soap or a can of milk at controlled prices from a state trading store only to resell the scarce item.	**5.** Explain what is happening here by referring to a supply-and-demand diagram.
Canadian and U.S. politicians are trying again this weekend to stabilize international wheat prices. The last international wheat agreement, shaped in 1967 in the Kennedy round of tariff negotiations, collapsed shortly afterward because price ranges were set too high to permit Canada and other wheat producers to get the returns they sought.	**6.** How could prices have been set too *high* to permit producers to get the returns they sought? Don't high prices mean high returns? Why would this cause the agreement to collapse?
Efforts to fix prices for many other commodities—potash, lead, zinc, copper, nickel, sugar, cotton, timber, jute—have also failed. The principle reason for such failures is the adequacy of supplies and availability of substitutes.	**7.** What does this say about the supply-and-demand curves for these products?

64 Part I *Microeconomics*

Proponents of the marketing board system cite the huge surpluses created in other countries, including the U.S., where the lack of a quota system and use of floor prices attracted thousands of farmers to try dairy farming. The market was soon flooded with surplus dairy products.

The surpluses became so huge that the U.S. government, sensitive to the farm vote, thought it was wise to buy the surpluses from the farmers and sell them on the open world market. It has also run a herd buy-out program, paying farmers to take 1.4 million cows out of production. Better to control production, proponents of quotas argue, than allow wasteful and expensive surpluses.

But then, the U.S. government could have decided to let the market take its course, leaving the dairy farming industry on its own to determine its own size.

8. Suppose Figure I-6.3 represents the supply-and-demand curves for milk.
   a) If a floor price is set at P_0, what will be the surplus?
   b) Under a marketing board, how much would be produced to achieve the price P_0?
   c) If the government bought the surplus in part a) above and sold it on the world market at price P_w, what would be the net cost?
   d) If there were no government intervention, what domestic output and import-export would occur? (Assume the world price remains at P_w.)

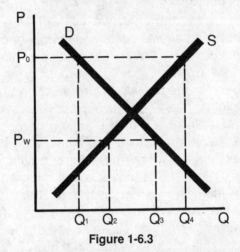

Figure 1-6.3

It cannot be doubted that in several key areas of social policy Ontario is in crisis: desperate shortages afflict health care, which takes a third of the budget; education, which takes a fifth; and housing and social services, which together take about another fifth. But every serious study of these problem areas has made the same point: what is needed, before all else, is structural reform, making greater use of those useful allocative devices, competition and the price system, to better deploy existing resources.

This in no way jeopardizes the commitment to social equity in these areas. But the Liberal government has not addressed the need for reform. Its answer is to try to float each crisis away on a raft of dollars.

9. a) What is the main reason for these "shortages"?
   b) What is meant by the phrase "to better deploy existing resources"?
   c) Give two examples of how using the price system could better deploy existing resources.

I-6 Government Fixed Prices 65

I-7 At the Margin

One of the basic principles of microeconomics is that decisions should be made at the margin. This means looking at the costs and benefits of a small change in quantity and calculating the desirability of that change on the basis only of the extra costs and benefits involved, ignoring overall profitability or desirability of the operation as a whole. The adjective "marginal" is used to capture this idea of *extra* costs or *extra* benefits due to a unit quantity change. The change should be undertaken if marginal costs are less than marginal benefits, since more will be added to total benefits than will be added to total costs. When marginal costs exceed marginal benefits, quantity should be decreased. The "right" quantity is therefore the quantity at which marginal costs equal marginal benefits. Marginal analysis, then, consists of finding the quantity for which marginal costs equal marginal benefits.

Why can the overall profitability or desirability be ignored when contemplating these changes? The reason is that by equating marginal costs and marginal benefits, we will automatically be maximizing the difference between overall costs and overall benefits. This is not to say that the difference between overall benefits and overall costs will necessarily be positive; once we have moved to a position in which marginal cost equals marginal benefit, the overall picture must be checked to ensure that overall benefits exceed overall costs.

This procedure of equating marginal benefits and marginal costs is a valuable tool in any economic analysis involving maximization. The typical example is a firm maximizing profits: Profits are maximized when the firm produces the output at which marginal cost equals marginal benefit, where in this instance marginal benefit is more appropriately called marginal revenue. Another example is the consumer maximizing utility: Utility is maximized by buying the quantity of the good at which its marginal cost (its price) is equal to its marginal utility.

The most interesting cases of application of this marginal principle, however, occur when conflicting goals are being maximized by business and government. Business wishes to maximize its profit and will therefore equate its marginal cost with its marginal benefit. Government wishes to maximize society's overall welfare and will therefore want business to equate society's marginal cost with society's marginal benefit. The problem is that the firm's marginal cost isn't necessarily the same as the marginal cost to society and the firm's marginal benefit isn't necessarily the same as the marginal benefit to society. For example, marginal costs may differ because of external diseconomies: Pollution costs may be borne by society and not by the firm. As another example, marginal benefits may differ because the firm has little competition and faces a downward sloping demand curve. Society's marginal benefit is measured by the price, but the firm's marginal revenue is measured by the price *less* the loss in revenue on existing output due to the fall in price caused by the sale of the extra unit of output. In examples such as these the government feels justified in taking policy action to adjust the operation of the market mechanism. These cases will be made clearer and illustrated in later sections of this book.

If external economies and diseconomies* are ignored, the relevant marginal benefits and marginal costs to society are given by the demand and supply curves, respectively. This is so by the definitions of the demand and supply curves. At a given price the maximum amount demanded is that quantity at which price equals marginal utility: At a smaller quantity, marginal utility would exceed price and the consumer would gain by demanding more, and at a larger quantity marginal utility would be less than price and the consumer would gain by demanding less. At a given price the maximum amount a firm will produce is that quantity at which marginal cost equals that price—it could increase its profits by changing quantity if that were not the case.

I-7A Cross-Subsidization

Hiding the cost

THE NATIONAL Transportation Agency will continue its inquiry into whether Via Rail should be permitted to discount fares on off-peak days on the Montreal-Toronto route. Why, in a market economy, are the complaints of Via's competitor, Voyageur bus lines, even given a hearing? Because, among other reasons, Voyageur argues such margin-shaving in the corridor would prevent it from cross-subsidizing less-profitable routes through its fare structure.

Call-Net Telecommunications Ltd. has just been ordered by the authorities to cease offering long-distance telephone service in competition with the Telecom Canada monopoly, headed by Bell Canada. Why? Because, it is said, competition on long-distance rates would prevent Bell from cross-subsidizing local service.

Canada Post appears to have weathered yet another strike. This time service was unaffected, but on other occasions it has been interrupted for weeks. Competition would improve service and make strikes much less disruptive. So why must we submit to Canada Post's statutory monopoly? Because, the corporation and its unions agree, competition for city mail would prevent Canada Post from cross-subsidizing rural service via uniform rates.

Cross-subsidizing is unfair and inefficient

Cross-subsidization is an inefficient and unfair means to an even less legitimate end. What is its effect? The allocation of resources society prefers is measured by the price at the margin we are willing to pay as consumers. Divorcing price from marginal cost encourages greater consumption of high-cost goods and services and less of low-cost than we would find ideal.

* External diseconomies are costs, such as pollution, that are not levied on producers. External economies, such as smells from a bakery, are benefits for which the producers are unable to charge.

Furthermore, where there are economic, social, or even political reasons for subsidization, burying the subsidy in the price system is the worst way to go about it. The first reason is simply that it is hidden: the cost of the subsidy is unknown. Second, it is doubly distorting: to push prices too low in one area, they are held too high in another.

Note: Suggested answers to the questions below appear in the Appendix.

1. Explain, in terms of marginal cost and marginal revenue, why Via Rail would want to discount fares on off-peak days.
2. In light of your answer to part a), why is this argument not true for Voyageur as well?
3. What is inefficient about cross-subsidization? Hint: Explain in terms of society's marginal benefits and marginal costs.
4. Explain what is meant by the statement "The allocation of resources society prefers is measured by the price at the margin we are willing to pay as consumers," as stated in the fourth paragraph.
5. Use supply-and-demand diagrams to show what is meant by the last sentence of the second-last paragraph.
6. What does it mean to say a price is too high or too low, as stated in the last sentence? What price is right?

I-7B Electric Metering

Time of Day Metering Near

By Edward Clifford

The advent of "time-of-day" electric metering, particularly for industrial and commercial power customers, may not be far away, according to the experts.

Such meters not only measure the amount of electricity used by a customer, but when it is used. The customer is subsequently charged a higher amount for power consumed during high-demand hours.

If this system is practiced widely, and providing the price difference between peak power and off-peak power is significant, power consumption can remain at a relatively even level instead of altering between highs and lows, on both a daily and seasonal basis.

This means less generating capacity needs to be built to meet peak demand, and less fossil fuel is required to fire up thermal stations.

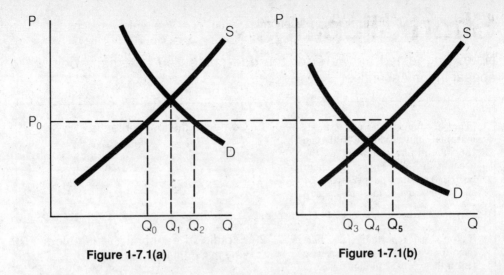

Figure I-7.1(a) Figure I-7.1(b)

Suppose the current situation can be depicted as in Figure I-7.1(a), representing supply (marginal cost) and demand (marginal benefit) in peak periods, and Figure I-7.1(b), representing supply and demand in off-peak periods. The supply curve is the same in both periods because the same generating facilities are used. The current price is a uniform price P_0 for both peak and off-peak electricity.

1. What statement in the article suggests that the marginal cost of electricity is rising, and is probably above average cost during peak hours?
2. From the diagram, what quantity of electricity will be generated in each period?
3. Does marginal cost equal marginal benefit in each period? If not, use the diagram to show whether marginal cost exceeds marginal benefit or vice-versa.
4. Assuming that the demand for off-peak power is not affected by the price for peak power and that the demand for peak power is not affected by the price of off-peak power, explain clearly how society can benefit from charging different prices for peak versus off-peak power. What specific solution would you suggest, in terms of the diagrams?
5. How would you measure the net gain to society? Be explicit, by explaining how this gain can be measured on the diagrams.
6. The two diagrams are not independent, since a change in the price of one kind of power will affect the demand for both types of power. How would this phenomenon best be captured in the diagrams? How would recognition of this affect your answer to question 4 above?

Short Clippings

Note: Suggested answers for the odd-numbered questions below appear in the Appendix.

[The] economist who understands marginal analysis has a "full-time job in undoing the work of the accountant." This is so, Alexander holds, because the practices of accountants and of most businesses. . . .

1. Complete the last sentence.

Exxon Corp. reportedly just chartered for $900,000 a year a Japanese tanker that costs $2 million a year to operate. One reason for the seemingly illogical deal: Japanese union rules require the owner to pay its seagoing crew its full wages whether or not the ship is employed.

2. Explain in economic terminology why this deal is not illogical.

The capital-intensive nature of the business—the high fixed-equipment costs in barns, heating and feeding units—forces the producer to sometimes operate to reduce losses rather than to make a profit.

3. Explain what is meant by "sometimes operate to reduce losses rather than to make a profit." What economic terminology is relevant here?

Continental Air Lines, Inc. last year filled only half the available seats on its Boeing 707 jet flights, a record some 15 percentage points worse than the national average.

By eliminating just a few runs—less than five per cent—Continental could have raised its average load considerably. Some of its flights frequently carry as few as 30 passengers on the 120-seat plane. But the improved load factor would have meant reduced profits.

4. Explain in economic terminology how improving the load factor would have meant reduced profits.

Instead of just sending some workers home earlier, as Kentucky Fried Chicken is doing, some companies may actually close shop earlier. "With the minimum wage going up, hours that once were marginally profitable might become unprofitable," says John Toby, a vice president of Jerrico, which is considering closing earlier.

5. What rule would a profit-maximizing firm employ to determine how early to close when faced with a higher minimum wage?

---

70 Part I *Microeconomics*

There are times, though, when the decisions dictated by the most expert marginal analysis seem silly at best, and downright costly at worst. For example, Continental will have two planes converging at the same time on Municipal Airport in Kansas City, when the new schedules take effect.

This is expensive because, normally, Continental doesn't have the facilities in K.C. to service two planes at once; the line will have to lease an extra fuel truck and hire three new hands—at a total monthly cost of $1,800.

6. Explain how marginal analysis could justify this case of very high marginal cost.

While the international rates appear to be well above the fully allocated costs of air cargo, the domestic rates seem to cover marginally, if at all, the cost of air cargo.

7. If you were the president of Air Canada, what would be your reaction to the information in this clipping about your airline?

The phone companies are at it again—trying to persuade us that the current charges for local phone service are somehow unfair. The companies want to do away with flat-rate local phone service and replace it with a scheme that would turn our home phones into pay phones.

Every local call would be charged. The cost could vary according to the length of the call, or the time of day, or the distance of the other party, or any combination of the above.

8. a) Where are we in Figure I-7.2 under the current system? What is the marginal utility of the last local phone call made?
b) What price would a profit maximizing firm like to charge per local phone call? What would their profit or loss be?
c) If the phone company were allowed to charge a price so that it could just cover its costs, what would the price be and how many phone calls would be made?
d) What price would an economist recommend? Show the resulting profit or loss.
e) As an economist, what argument would you make against the prices suggested in 8b) and 8c)? What makes your price better?

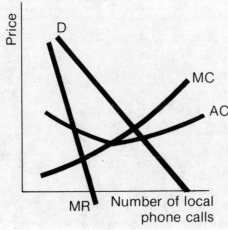

Figure 1-7.2

One of the aims of the study is to try to get users off peak-period times, when power is more expensive to produce. The preliminary findings also suggest time-of-day metering for the large user. This part of the study says that most users come out with lower power costs by careful planning of power use.

The objective of the new system of rates would be to try to redistribute the periods of heaviest demand for electricity, both during the day and the year. Hydro would be delighted, for instance, if half the electricity users were to use it at night rather than during the day.

As far as Hydro is concerned, he said, waste is "any price below what it costs you to produce another unit of electricity, or what you save if you produce one less unit of electricity ... That is, any price below marginal cost ... leads to wasteful use of electricity."

The Hydro studies team is conducting a cost-benefit analysis of the feasibility of abandoning bulk metering such as is now used in multi-family dwelling units, since such a move would yield significant conservation benefits. It was discovered that residents in bulk-metered apartments use on average 34.5% more electricity than when individually metered.

9. a) What does this clipping imply about the nature of Hydro's cost curves at the current output level?
b) By means of a diagram, explain why Hydro would be delighted if half the users used it at night.

10. a) Show on a diagram the wasteful use of electricity discussed in the first paragraph.
b) Does the 34.5 per cent difference in elasticity use support or contradict Hydro's argument in the first paragraph? Explain.

I-8 Taxes and Subsidies

No one denies that the price system—the use of prices to allocate and distribute scarce goods and services—has flaws. There are three major flaws that receive considerable attention from policy makers. The first is that sometimes the dynamics of the operation of the price system create unnecessary (and undesirable) cycles. The second is that the income distribution created by the system is thought by many to be "unjust." The third is that often not all costs are borne by the suppliers (external diseconomies) or not all benefits are paid for by the consumers (external economies).

A variety of government policies have been directed at these problems. Examples are price floors, price ceilings, marketing boards, fixed prices, taxes, subsidies, and various kinds of income redistribution schemes. Of all the interferences in the market mechanism that have been proposed by governments, the use of taxes and subsidies is thought by economists to be the least unpalatable. This is because the use of taxes or subsidies still allows allocation and distribution to be accomplished by the price system; all the taxes and subsidies do is alter the prices to what the policy makers feel are the "correct" prices.

An example often used to illustrate this is the case of pollution, an external diseconomy. Pollution is a "diseconomy" or cost to society since it detracts from our well-being. This cost, however, is "external" to the firm doing the polluting: The firm does not bear the cost of the pollution. Thus the cost of producing the good in question consists of the normal input costs, borne by the firm, plus the cost of the pollution, borne by society. Because the firm does not bear all the costs associated with production of the good, the profit-maximizing quantity of the firm will involve a net loss to society in the form of excessive pollution. By taxing the firm's output, the government can force the firm to internalize the pollution cost and thereby cause them to change their profit-maximizing output level, eliminating this net loss. A more explicit example should make this clearer.

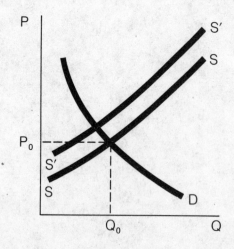

Figure 1-8.1

74 Part I *Microeconomics*

Suppose the supply and demand curves for this good can be shown as in Figure I-8.1. But suppose that firms pollute the air during their production process. To society this is a cost, but this cost is not included in the supply curve, since the supply curve reflects only costs internal to the firm. Thus the supply curve does not incorporate all of the costs to society of producing that good. At the quantity Q_0 the price P_0 reflects the benefit to society of consuming the last unit of output (i.e., from the meaning of the demand curve), but it does not reflect the total extra cost to society of producing that last unit of output, since the pollution cost is ignored. If the government imposes a tax on the firm's output, shifting the supply curve to what it should be (S'S'), from society's point of view, the resulting price and output will be "correct." Here "correct" means that the total extra cost to society of producing the last unit of output is equal to what society is willing to pay for that unit of output (the benefit to society of that last unit). If this equality did not hold, society could gain by changing the output level. If, for example, the cost of producing the last unit of output were greater than the benefit derived therefrom, society could gain by reducing output (costs would be reduced by more than benefits). If the cost of producing the last unit were less than the benefit derived therefrom, society could gain by increasing output (costs would be increased by less than benefits).

This illustrates a general use of taxes and subsidies—to force incorporation of external economies and diseconomies into the supply and demand curves, leading to the "correct" output levels. The pollution example given above is incomplete since it does not recognize the possibility of applying pollution-control devices; the extension of the analysis to include this possibility uses the same principle of applying taxes to make the market move by itself to the correct output level.

Because many different kinds of taxes and subsidies could be applied in a given instance (lump-sum taxes, per unit taxes, ad valorem taxes, and taxes on profits, for example) economic analysis must determine the most appropriate kind of tax or subsidy as well as the appropriate magnitude. The articles in this section allow examination of this problem.

I-8A Pollution

Pollution charge

Opposition leader Bob Skelly wants to charge for the right to pollute.

The present system of allowing holders of pollution control permits to dump wastes without paying for the privilege means that the cheapest course of action for polluters is usually also the most costly to the environment, Skelly said.

The government should charge for the right to pollute, to make environmentally acceptable solutions the cheapest solutions to waste disposal problems, he said.

"For this province to underprice life sustaining resources is criminal," Skelly said during the first day of debate on Environment Minister Stephen Rogers' spending estimates.

In a market economy, the consequence of underpricing a resource is that the resource gets wasted, he said.

Rogers told Skelly he thinks his suggestion is a good one but said his ministry's highest priority right now is dealing with the handling of toxic wastes.

Outside the house, Rogers added that he doesn't plan to implement Skelly's suggestion immediately because it would drastically change the economics of existing businesses.

Note: Suggested answers to the questions below appear in the Appendix.

1. If a charge were levied to cover the cost of waste disposal, show the effect it would have on the cost curves of polluting firms. How would it affect the market for the polluting firm's product? How would this lead to less pollution?

2. If the government were to charge for the right to pollute, as suggested in the second paragraph, how should it determine the appropriate price to charge?

3. A judicious choice of quotas under the current system and the proposal (of charging for the right to pollute) are equivalent in the sense that they would both create the same amount of pollution. Which of these two methods would you recommend? Why?

4. A variant of the quota system is to permit polluters to buy and sell these quotas. Comment on the relationship between this method and the suggestion of charging for the right to pollute.

I-8B Day Care

Wrong emphasis in day-care policy

SUDDENLY, EVERYONE'S a supply-sider. The critique of Keynesian-style expansion of the economy has always been that stimulating aggregate demand is of little use if aggregate supply cannot rise in response.

When it comes to the government's new $5.4-billion day-care plan, however, the Keynesians in Opposition have undergone a miraculous mass conversion: what's the point of boosting demand through tax breaks for parents, they now ask, if the supply isn't there? Instead, they would boost supply through state funding of day-care centres, whether the demand is there or not.

The shock horror statistic of two million children and only 200,000 licensed spaces is misleading. One could as well say 10 million drivers, 30,000 Mazdas. The parents of those 1.8 million other children are clearly not all clamoring to get their offspring into licensed centres, and it is dishonest to suggest that they are.

But let us accept for argument's sake that there is such an imbalance. Why isn't the demand being met? Why are there so few spaces? Is some structural failure in the market preventing day-care operators from increasing supply? No. The problem lies in a lack of effective *demand*. Many parents can't afford day care, which in turn prevents them from taking a job so they could afford it. Why don't day-care operators drop their prices then? Because they're too low as it is. They cannot now pay enough to attract and hold skilled workers; high turnover reduces the quality of care, which in turn affects demand.

A demand-side problem calls for a demand-side solution. Subsidizing parents raises the demand for day care at any given price. This in turn drives up prices, calling forth greater supply—that's *more spaces* — thus bidding up wages of day-care workers, so reducing turnover and improving quality. The superiority of this over supply-side solutions is that it gives control over funding to parents: their choices then determine what type and amount of day care will be provided. If $5 billion will create a certain number of spaces, it will do so whether it is allocated by politicians or by parents. Politicians prefer direct grants to centres—lots of lovely sod-turning ceremonies and photo opportunities. Parents prefer choice.

---

1. Explain how the main ideas of this editorial can be captured in a supply-and-demand diagram.
2. How would you answer the question posed in the second paragraph?
3. Under what circumstance would 10 million drivers all want a Mazda? Explain how this relates to the "shock horror" statistic in the third paragraph.
4. What is the difference between demand and effective demand? (fourth paragraph)
5. What is the external economy or diseconomy here that prompts government interference?

I-8C Government Cafeterias

Bouey rapped on meals deal

OTTAWA (CP)—A Conservative MP has criticized Bank of Canada Governor Gerald Bouey for giving the bank's employees a $250-a-year raise to compensate them for the loss of subsidized meals in the bank's cafeterias.

The bank doubled its meal prices April 1, John MacDougall, MP for the Ontario riding of Timiskaming, said in a news release Friday. "But at the same time, all employees were given a $250 raise to compensate.

"This clearly defeats the whole purpose of raising the prices in the first place," he said. "In effect, any savings from an increase in cafeteria prices is going in one pocket and out the other.

"I wish to remind the governor that it is not up to the people of Canada to ensure that Bank of Canada employees are well fed," MacDougall said.

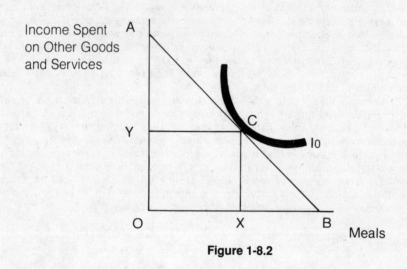

Figure I-8.2

1. It is not unusual for employees to take benefits such as medical, dental or subsidized meals in lieu of wage or salary increases. The article above deals with a case where the employees have had a benefit or subsidy removed and have been compensated by a wage increase. In order to analyze the events in this article an "indifference curve" diagram can be used. In Figure I-8.2 the line AB represents the consumer's budget constraint with the subsidy so that he or she purchases OX of meals and OY of other goods and services.

   a) Show the impact of the removal of the subsidy without compensation. Clearly show and indicate how this affects the consumer's purchases of meals.
   b) How can the technique used in part a) be used to derive the demand curve for meals?

c) If the wage increase compensated this individual just enough to keep the level of satisfaction unchanged, would he or she make the same purchases of meals? If not, how much would be consumed? Explain. (Hint: Shift your new budget line in a) out to the point at which it just touches the old indifference curve.)

d) Will the Bank of Canada save money or spend more with the wage increase granted in c)? Show the savings or additional costs incurred.

e) In the third paragraph the indications are that the costs will be the same, that is, from the individual point of view the compensation is just enough to make the same level of purchases (point C) as with the subsidy. Show this by drawing a new budget line that allows them to purchase OX meals but at the new prices. Will the bank employees still maintain the old level of satisfaction? If not, show what their new level of satisfaction is.

f) Use your answers to d) and c) to comment on the first sentence of the third paragraph.

Short Clippings

Note: Suggested answers for the odd-numbered questions below appear in the Appendix.

A. The Effects of Taxes and Subsidies

Asked if the two-cent levy would be reflected in the price of gas at the pump, Carney said: "The market situation will dictate whether the price rises."

1. How will the market dictate how much the price rises? Refer to a supply-and-demand diagram in your answer.

Thorpe agreed. "The Canadian Dairy Commission feels that the subsidy, while paid to producers, is considered a subsidy to consumers," he said. "If the producer didn't get that subsidy, the consumer would have to pay a higher price for dairy products."

2. a) Does it make any difference if the subsidy is given to producers or to consumers? Explain.
   b) Under what circumstance will the subsidy benefit only the consumers?

Esso, a division of Imperial Oil Ltd., said in a release that the "ability of a retailer to recover the tax increase through an increase in retail prices is not certain, and will depend on market conditions in any given area."

3. Exactly what market conditions will allow a retailer to recover the tax increase? Illustrate on a supply-and-demand diagram.

Chevron spokesman Roy Jolly said prices were expected to come into line with the new taxes, but added that "the market is a very strong force in regulating prices."

4. Explain the circumstance in which prices will "come into line with the new taxes."

It has been an article of faith for many left-wingers that rent subsidy programs will inevitably fail to benefit their recipients. Landlords will simply raise rents and pocket the subsidies, they contend.

I will start taking that argument seriously the day I see a union pass up a wage increase on the ground that its members' landlords would just raise their rents and take it all away.

5. Explain why it will be impossible for landlords to simply "raise rents and pocket the subsidies." Will they be able to pocket *some* of the subsidies? Explain why or why not.

Robyn Allan, an economist at B.C. Central, said the estimate of $500 million flowing to the U.S. Treasury is based on a maximum tariff of 27 per cent on softwood lumber exports to the U.S.

"We took 27 per cent of a smaller volume at a higher price," she said. "And we came up with close to the $500 million figure."

6. Explain why the 27 percent was applied to "a smaller volume at a higher price.".

80 Part I *Microeconomics*

The controversial tax on marine fuel originally caused prices to increase 15 per cent and led to us selling 40 percent of what we had sold, even though the number of ships stayed the same. With the removal of the tax, the current situation is not going to change for a few months as the oil companies have gradually stopped supplying the marine market. However if the tax savings are passed on, it will make us more competitive and supply and demand will increase.

7. a) What is the price elasticity of marine fuel?
   b) Under what circumstances would the tax savings be passed on?
   c) There appears to be a price decrease here that is resulting in an increase in *both* supply *and* demand. How can this be?

B. Pollution

Environmental pollution is violence, he pointed out. In terms of cost to our health, safety, and property damage, it "absolutely dwarfs crime in the streets and disorders on the campus in terms of numbers of people affected, seriousness of impact, disastrous effects and property damage."

All the riots this past year did $300 million property damage, he noted. "Our air pollution alone cost us $11 billion a year."

1. The implication of these statements is that pollution should be greatly reduced, if not eliminated, or that too much money is being poured into fighting crime relative to that used to fight pollution.
   a) Why is it that these conclusions cannot be drawn from the facts cited?
   b) What facts would be needed to support those conclusions?

Aside from the fact that non-returnable bottles and cans are wasteful, they are problems because they add to a mushrooming volume of garbage that big cities are finding increasingly difficult to dispose of. They also are a major cause of litter, and broken bottles and cans may lead to nasty accidents.

The best answer would seem to be to phase out the use of non-refillable bottles and to impose a special tax on pop cans. Beer drinkers already have to pay more if they want cans.

2. The environmentalists complain because producers of non-returnables do not bear the cost of the externalities (garbage and litter) caused by their product. Imposing a tax on the non-returnables is designed to raise their money cost to cover the total (private plus social) cost. Banning them is an alternative solution. The first solution creates an "optimal" amount of pollution (garbage and litter) whereas the second solution creates zero garbage and litter, a less than "optimal" amount.
   a) What is the meaning of the term "optimal" amount of pollution?
   b) Using a supply-and-demand diagram, show how the "optimal" amount of non-refillable bottles can be determined. Assume that you are able to measure the social cost of pollution at all output levels.

Part I   *Microeconomics*

c) In terms of your diagram, what is gained by adopting the tax solution rather than the banning solution? Or rather than a "do-nothing" solution? Couch your answer in terms of consumers' and producers' surpluses.

It would be a mistake to start counting the profits from power exports without first adding up all the associated costs. Those include the environmental costs, whether the loss of a fertile valley to a hydroelectric dam or pollution from a coal plant.

3. How would a government go about ensuring that these costs are properly accounted for?

Federal Environment Minister Tom McMillan says: "Just as businesses have to meet business codes and labor laws so also are they realizing the environmental costs have to be factored in. If a company cannot survive unless it transfers its environmental costs to somebody else, maybe it ought not to survive. That's what the marketplace is all about."

4. a) How can the government ensure that a business factors in environmental costs?
   b) Give an example of how a company "transfers its environmental costs to somebody else." What terminology do economists use to refer to this?

"We have been able to produce nickel in the province and we have been able to do so by having a byproduct—acid rain—which is damaging to another industry such as the forest industry, or the commercial fishing industry or tourism." For Canada's companies, the portent of environment laws is clear. Turpin concludes: "They may not be economic for the company, but they are very cost-effective for society."

5. a) What terminology do economists use to refer to damaging byproducts?
   b) Explain what is meant by saying that environmental laws are "very cost-effective for society"? Hint: See the following clipping.

McMillan and Ontario's Environment Minister Jim Bradley warn that polluters must pay to clean up their act. As biology professor David Turpin of Queen's University in Kingston, Ont., says: "If governments pay, it obliterates the value of competition. But if everyone is told to clean up, then the company that does it for the least amount of money will have the least expensive product and will prosper in the marketplace."

6. a) What is the most appropriate way for the government to make polluters "clean up their act"?
   b) Does this imply zero pollution? If not, why not?

SASKATCHEWAN has lifted its 15-year ban on the sale of beer and soft drinks in cans. Pressure from consumer groups appears to have been the driving force for recent legislation overturning the 1972 ruling against canned beverages.

The former NDP government banned cans in 1972 as an uncontrollable threat to the environment.

7. Show on a supply-and-demand diagram where the pressure from consumer groups is coming from. Be sure to build into your diagram consumer environmental concern.

C. Miscellanea

Far from imposing a penalty, Canada continues to encourage oil imports with a subsidy from the public treasury.

This makes our measured rate of inflation look better than it is. But it doesn't really save any money. The oil-producing countries from which we import get their pound of flesh—and all the more because of our wastrel approach to energy policy.

1. Explain why it is suggested that Canada's policy gives an even bigger pound of flesh to the oil-producing countries.

Energy costs are hitting the poorest people in society twice as hard as the wealthiest and the situation is likely to get worse as fuel prices rise according to a study by the Social Planning Council of Metropolitan Toronto.

2. Compare the advantages and disadvantages of the following two policies for dealing with this problem:
   a) Place a freeze on fuel prices.
   b) Give cash to the poor in an amount sufficient to make up the difference in fuel expenditure resulting from higher fuel prices.

Patterson said it is "gross oversimplification" to assume the removal of rent controls alone will create any significant amount of new construction.

"If this were true, it would not be necessary for Alberta to put in half a billion dollars to subsidize the rental housing market at the same time rent controls are removed."

3. a) By what mechanism is the removal of rent controls supposed to lead to an increase in new construction?
   b) Suppose that you believe that the mechanism in part a) works. How would you explain why Alberta is planning to subsidize the rental housing market?

"Almost all the calls I get are for numbers in the phone book," says Pam, who estimates she talks to 600 callers each three-hour work split. "I try to be diplomatic and say, 'the number is listed as . . . ,' but what I really mean is, 'next time look in your book.'"

Callers may start looking in the book because soon it's going to cost 20 cents to get the number from Pam.

4. Would an economist condemn this action because it unjustly raises prices, or praise it because it justly raises prices? Explain your reasoning.

The suggestion that British Columbians should pay a deterrent fee at the doctor's office will only further undermine the medicare system the doctors say they are afraid is slipping.

But it would approach the perceived problem—and that's all it is at the moment—from the wrong perspective. If there is a problem of over-use of medical services, that surely is a medical problem, not an economic one.

5. The problem, as expressed in the last sentence, is over-use of medical services.
   a) How would an economist define over-use?
   b) What solution to this problem would an economist prescribe? Explain the economic reasoning involved.

The universal access to medical care has led to non-essential aspects which are costing Canadians billions of dollars more than is necessary. The move to user fees would be a "healthy" step even in the face of the arguments about denying low income families access and financially ruining people. The proposals of setting a small charge, amounting from 5 to 10 per cent of the actual cost, include income compensation for low income families and put the onus on them to economize.

6. a) By referring to Marginal Utility theory explain what the current consumption of health care is in terms of marginal utility.
   b) If the user fee is introduced what will happen to the consumption of medical care and other goods?
   c) Why might an economist prefer a user fee?
   d) If we give "low-income families" income compensation so they can afford the same level of medical care as before the user fee, why would they economize?

"The committee makes no recommendations with respect to taxing such necessities as food.

"The committee recommends, however, that if necessities are taxed, it should be only on condition that low and lower-middle income groups are fully and immediately compensated for the incremental burden they bear, and that such compensation is fully indexed."

7. Use Figure I-8.2 (change meals to necessities and food) to show that the consumer will be better off if fully compensated for the tax imposed. Hint: Assume that "fully compensated" means that they are given sufficient extra income to maintain their pre-tax consumption combination.

The result: the profusion of Rolls-Royces in Britain, which is generally seen as a sign of prosperity, is nothing of the sort. It is a sign of a stupefyingly high level of taxation and a tax system that deters people from using their money productively.

8. Explain how a very high marginal tax rate would lead to this behavior. Hint: Consider the opportunity cost.

As the subsidies diminish, land prices will crash. As always, everywhere, the main effect of agricultural subsidies here has been to push up land prices to uneconomic levels.

9. Explain how and why land prices change as subsidies change.

I-9 Foreign Competition

Because Canada is a small, open economy, its markets are kept competitive mainly by foreign competition. Although this ensures that Canadians are able to buy from the world's lowest-cost producers, it also means that domestic industries are limited to those in which Canada has a definite comparative advantage. Not wanting to be a nation of "hewers of wood and drawers of water," Canada has encouraged development of manufacturing industries, often protecting them with tariffs or quotas.

This protection, in the face of economists' advocacy of free trade, has been rationalized in several ways. First is the "infant industry" argument: To get an industry started and functioning efficiently requires a few years of protection from competition; once on its feet the protective trade barriers can be removed. Second is the "vital industry" argument: Although it may be cheaper to buy a good elsewhere, in the event of a war or some international emergency, an assured supply of that good may be thought necessary. Third is the "nationalism" and "prestige" argument: All developed countries, some people think, should have certain industries operating in their economy either for "prestige" or to enhance their "nationalism." Fourth is the "uncertainty" argument: Foreign suppliers may be unreliable, implying large fluctuations in quantities delivered and prices charged, disrupting domestic industry using this good as an input. Fifth is the "foreign monopoly" argument: It may be the case that foreign suppliers may act as monopolists (or as a cartel) in the absence of domestic competition, forcing Canadians to pay a monopoly premium. Sixth is the "diverse economy" argument: Canadians may perceive distinct social advantages (e.g., giving citizens a wider choice of occupations) and stability advantages to having an economy characterized by a wide variety of industries, rather than being reliant on the fortunes of and opportunities provided by a small number of industries.

In response to these arguments, the Canadian government has imposed some tariffs and quotas to protect certain Canadian industries. At the same time, however, the Canadian government recognizes the benefits of trade, and in negotiations with other trading countries is constantly pressing for lower tariffs imposed by others and is in turn being pressured to lower its own tariffs. Lowering Canadian tariffs accomplishes several objectives. First, it serves to increase trade and the benefits derived therefrom. Second, it keeps domestic industries competitive, ensuring fair price for Canadian consumers and acting as an anti-inflation force. And third, it acts as a negotiating instrument in obtaining reductions in foreign tariffs, aiding Canadian exporters.

Thus although the Canadian government protects Canadian industries for the reasons given earlier, it keeps that protection at the lowest level it thinks is feasible. As a result, industries are constantly complaining that their protection is insufficient.

I-9A **Lumber**

Trading day busy as higher lumber prices anticipated

By ROD NUTT
Sun Business Reporter

U.S. lumber-buyers said Friday their trading day was busier than usual in anticipation of higher prices Monday.

But traders in Boston, New York and Chicago said supply and demand would ultimately determine where prices settled and doubted Canadian mills would be able to make a 15 per cent price hike stick to offset a U.S.-imposed tariff.

Bernie Futter of Futter & Co. Lumber Corp., Long Island, N.Y., predicted prices would rise "$10 to $15 US a thousand board feet Monday."

"Initially, prices will go up because U.S. mills will pass along an increase but the Canadian mills won't be able to go above those levels," Futter said.

"The Canadian industry has to realize it isn't a world unto itself . . . the market price and will ultimately be set by supply and demand and the Canadian mills will have to absorb at least half the increase caused by the 15 per cent tariff."

Futter predicted that lumber prices would eventually settle back to today's levels and that Canadian producers would have to "swallow all the increase" caused by the tariff.

Note: Suggested answers to the questions below appear in the Appendix.

1. What is meant by "supply and demand would ultimately determine where prices settled"? Illustrate on a supply-and-demand diagram.

2. Under what circumstances would the Canadians be able to make a 15 per cent price hike stick to offset the 15 per cent tariff?

3. Under what circumstances would the opposite be the case, as predicted in the last paragraph?

4. In terms of your supply-and-demand diagram, what relationship would have to hold between the supply-and-demand elasticities for the Canadian mills to be forced to absorb more than half of the tariff as stated in the second-last paragraph?

5. Interpret the fourth paragraph in terms of your supply-and-demand diagram. Under what circumstances would the Canadian mills be able to "go above these levels"?

I-9B Apples

Ottawa slaps import duties on U.S. apples

Canadian Press
OTTAWA

Revenue Canada has imposed import duties on U.S. delicious apples—the kind with distinctive bumps on the bottom—that Canadian growers say have been dumped on the Canadian market at unfair prices.

The preliminary duties—a kind of border tax—were the result of an investigation that found the apples were being exported to Canada at prices as much as 67 per cent less than the cost of production in the United States. The dumping has resulted in lower prices for Canadian apples.

The duties won't be made permanent until the Canadian Import Tribunal conducts its own inquiry to determine how much dumping has hurt Canadian growers.

The duties collected will vary with different varieties of delicious apples and how much each particular shipment of U.S. apples falls below the price Revenue Canada feels should be a fair level.

Enough duties will be collected on each box of apples to wipe out the U.S. price advantage, which Revenue Canada calculates at an average of about 32 per cent.

The complaint was launched by the Canadian Horticultural Council on behalf of apple growers mainly in British Columbia and Ontario.

A spokesman said the problem is the result of a bumper apple crop in Washington state, which supplies about half of all apples imported to Canada.

---

1. What is meant by "dumping"?
2. What is meant by an "unfair price"? (first paragraph)
3. By using a supply-and-demand diagram, explain the impact of the Washington dumping on Canadian apple growers.
4. What effect does the dumping duty have on the diagram in question 3?
5. Explain why dumping can be profit-maximizing behavior on the part of the Washington producers, i.e., why would Washington growers sell their apples at prices as much as 67 per cent below their production cost?

I-9C Tariffs vs Quotas

Quota Victims Poor

The tariff or its equivalent is alive and well in Canada despite talk of free trade and condemnation of protectionism by others.

The favored method of protection these days is the quota. It doesn't sound quite so vile as a tariff but the effect is often the same. Canada has quotas on footwear, textiles, and automobiles. But that doesn't stop Canadians and their governments from screaming about protectionist gestures made by other countries, especially the United States.

However, it is the less powerful and poorer nations against which Canada likes to impose quotas. When they complain, they don't have the clout of the U.S. The victims are frequently Third World countries struggling to keep above the economic tidal waves often caused by the richer countries' incompetent management.

1. The author of this clipping is against trade protection and indicates that both tariffs and quotas have similar effects. Let us look at these two options, import quotas and import tariffs. Suppose the Canadian demand and supply of footwear, textiles or automobiles is given in Figure I-9.1.

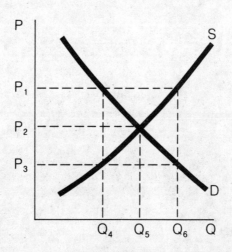

Figure 1-9.1

a) What combination of price and output do you feel best represents the current situation? Explain.
b) What is the current level of imports, as shown in the diagram?
c) What does the demand curve faced by importers look like? Refer specifically to the diagram in explaining how it would be derived.
d) Suppose a quota is imposed on imports, equal to half of the current level of imports. Draw a new diagram showing the new price and output levels. Explain how you obtained your answer.
e) When a quota is imposed there will be "quota profits." What are quota profits and to whom do they accrue? Explain how they would be measured in your

diagram from part d). Can you think of a way by which the government could reap these quota profits for itself?

f) Suppose an ad valorem import duty (tariff) were imposed. (Such a tariff consists of a percentage of the value of the import. An ad valorem tax of 30 per cent means that if the imports originally sold for $5, a tax of $1.50 would be payable.) Draw a diagram showing the new domestic price and output levels.

g) Explain how the duty collected would be measured in your diagram for part f).

h) Under which system (quotas or tariffs) would the domestic producers benefit most (or lose least) in the event of
  i) an exogenous increase in demand?
  ii) an exogenous decrease in demand?
  iii) an exogenous increase in the world price?
  iv) an exogenous decrease in the world price?

2. What is implied about the demand for our products by the U.S. and Third World countries that would lead Canada to pick only on the poorer nations? Aren't the Third World populations a bigger market for our products?

3. One of the arguments for protection is to allow domestic industries a chance to become competitive with foreign producers. What does the writer feel is happening because of protective measures?

Short Clippings

Note: Suggested answers for the odd-numbered questions below appear in the Appendix.

A. Tariffs

"If U.S. consumers are very insensitive to higher lumber prices, then prices could rise by nearly the full amount of the tariff, that is, 27 per cent. U.S. producers would bear the full cost of the tariff and Canadian producers would feel little impact."	**1.** a) What is the first sentence supposing about the elasticity of demand for lumber? b) Explain the serious error in this quote. Hint: A single word is incorrect.
Of more importance to the company is the removal of a 25 per cent tariff on the synthetic material they bring in from the U.S. It will be removed over 10 years in 2.5 per cent per year increments, which means Fireflex's raw material will eventually cost 25 per cent less. Those savings, said Carter, will be passed on to the customer.	**2.** Under what circumstances would the savings be passed on to the consumer? Illustrate on a supply-and-demand diagram.
Bank economist Michel Lefebvre says Americans need Canadian lumber so much that any countervailing duty "would hardly affect the volume of U.S.-bound exports."	**3.** What does this say about elasticity? Explain by means of a supply-and-demand diagram.
That traders said Friday, is an indication that the 15 per cent duty is being split between the Canadian mills and their U.S. customers.	**4.** Use a supply-and-demand diagram to show why the duty will be split.
Riley said the market will ultimately decide the price and whether Canadian producers will be forced to absorb the tariff to remain competitive.	**5.** a) Use a supply-and-demand diagram to show how the market decides who absorbs the tariff. b) Under what circumstances would Canadian producers be forced to absorb all of the tariff?
"There are all kinds of people who will be hurt by this," he said in a telephone interview. "There's the U.S. house builders, various rail and transport operations, publishers, terminal operators, ports, stevedoring companies—all would see declining business if there's a tariff imposed on Canadian forest products."	**6.** Explain how the group of U.S. businesses listed will be affected by the tariff. (Hint: The list of businesses can be split into two groups, each affected differently.)

B. Quotas

A report commissioned by the association claims that import quotas on Japanese-made cars will mean higher prices for small vehicles in Canada.	1. Explain how these higher prices will come about.
Sigh. In yesterday's column I said that restrictions on shoe imports into the U.S. would cost American consumers at least $1.28 billion while saving an estimated 20,000 jobs. The estimate of the number of jobs that would be saved is actually somewhat higher—26,000. That means consumers are paying close to $50,000 per job saved, not $64,000, as I reported yesterday.	2. Explain the economic reasoning that leads to the calculation of the cost per job saved.
Mr. Crandall sums up the argument succinctly: "Given the scant evidence that these quotas are advancing the competitiveness of the U.S. automobile industry, their desirability turns on whether Americans (or Canadians) wish to pay large premiums on their cars in order to increase the employment of auto workers at wages far above the manufacturing average."	3. Present the economic theory that would justify this statement.
He calculates that the import restrictions kept 6,000 textile employees at work last year. The average worker earned $10,000. But his job cost taxpayers $32,959. In sheer economic terms, it would have been better to allow all 6,000 to go on unemployment insurance.	4. Explain how each job could have cost the taxpayers so much.
"What the Canadian consumer might save on lower prices for textile goods could be more than offset by lost wages and increased welfare costs," a summary of the study concluded.	5. This clipping comments on the effects of not imposing an import quota or tariff on textiles. a) Explain in your own words what this quotation means. b) How would you go about measuring these savings and costs so as to determine the validity of the "more than offset" assertion? Be certain to comment on the number of years over which these calculations should take place.

Despite this increased share of the domestic market, the apparel manufacturers told the board they want more. "We are looking for a situation where we can be assured of supplying 75 per cent of our own market . . . consumers are being overly selfish in seeking the unrestrained right to demand bargains at the expense of our economic system . . . there is a cost to being Canadian."

6. How would you argue against the position taken in this item?

He added that the consumer is not guaranteed big savings when he buys imported clothing. Orders for foreign-made clothing must be made on a large scale and five or six months in advance of the shipment date. The importer also takes more risks than if he buys from Canadian manufacturers.

The result is often that the clothes are out of style or out of season when they arrive and the importer, to cover the risk of that happening, regularly puts on a high mark-up.

"The consumer has to pay for the risk," Crutchley said. "I don't see where the consumer benefits. Import prices are not lower in the stores."

7. This argument, by a spokesman for the textile industry, is trying to defend the current system of quotas. Do you think his argument is an adequate explanation of why "import prices are not lower in the stores"? If not, how would you explain this fact?

The importers complain that the domestic manufacturing industry cannot yet adequately make up the difference between the import limits and Canada's needs, and the result has been a black market in import quotas, high prices and inadequate selection.

8. a) What is the cause of the high prices?
b) Explain what is happening here in the black market in import quotas.

The Japanese have learned to love the quota system. And why not? It functions like a classic cartel, and the result is that each car sold in America—imported and domestic—costs at least $1,000 more than it otherwise would.

9. By means of a supply-and-demand diagram, explain why quotas on Japanese car imports would lead to an increase in the price of all cars.

Just how much have import restrictions cost Canadian consumers? Probably an average $888 for each Japanese car and possibly as much as $739 for each domestic car sold in 1985, the report concludes.

10. Use a supply-and-demand diagram to explain how these costs came about.

The quotas will restrict Japanese automobile imports into Canada to 170,400, and this represents about 18 per cent of the anticipated new car market.

Before the government began to intervene, 25 per cent of new car buyers were opting for Japanese models. That seven per cent difference amounts to about 60,000 new Japanese cars which will not be purchased by Canadian consumers simply because they won't be allowed into the country.

11. a) Use a supply-and-demand diagram to illustrate the impact of the quotas on the Canadian automobile market.
b) The clipping suggests that there will be 60,000 Canadians who wanted to buy a Japanese car but who won't be able to. How will it be decided who gets the Japanese cars?

The object of the sugar program is to keep U.S. prices stable at 18 cents a pound, relying on an import quota to insulate the U.S. market from world prices, which are a lot cheaper at eight to nine cents a pound.

The report, completed in April, estimates that U.S. sugar prices would fall and world prices rise to about 11 cents a pound if the program were abolished. The difference between that price and the current U.S. price costs consumers about $3.2 billion each year, the report said.

12. a) Use a supply-and-demand diagram to illustrate the sugar program and how its removal would save U.S. consumers $3.2 billion.
b) Explain why the world price for sugar would rise if the U.S. import quotas were removed.

"The report presents hard economic evidence that restrictive quotas are counterproductive because they discriminate against a select group of Canadian businessmen, and they limit the consumers' freedom of choice while creating very few, tremendously expensive jobs in the Canadian auto industry," Robert Attrell, president of the Japanese automobile dealers association, said earlier.

13. a) Which select group of Canadian entrepreneurs do you think are being discriminated against?
b) How do restrictive quotas limit the consumers' freedom of choice in a market system?
c) Why are the jobs created in the Canadian auto industry few and expensive?

C. Free trade*

"It is the maxim of every prudent master of a family, never to attempt to make at home what it will cost him more to make than to buy. The tailor does not attempt to make his own shoes, but buys them of the shoemaker. The shoemaker does not attempt to make his own clothes, but employs a tailor . . . What is prudence in the conduct of every private family, can scarcely be folly in that of a great kingdom."—*Adam Smith, in 1776.*

1. a) What economic law does this reflect?
b) What is the payoff to obeying this law?

While it is true that many products produced for export are not usable to answer local needs, it may be that the best way people in the Third World can clothe themselves is to produce garments, sell them to the First World and use the proceeds to buy what they need. This notion of pursuing one's comparative advantage in production and trading to get other items one needs is one of the oldest and most important pieces of economic understanding.

2. a) What economic law does this reflect?
b) What is the payoff to obeying this law?

Could it be that we have discovered here an old truth? Employment can be increased by imposing duties on goods. There is a double advantage here. If the tariff moves production to Canada there is a reduction in unemployment and an increase in government revenues. If production is not moved here, the government still gets revenue from the customs duties. That sounds like an excellent way to solve our economic problems.

3. How would you rebut this suggestion?

*For more on free trade see Section II-11.

"As a result of free trade, lands have been devalued," Sperling said. "We know of one vineyard that cost the grower about $35,000 an acre—and the land isn't worth $10,000 an acre at the moment."

4. Explain how free trade could affect the value of land.

Protectionism does not produce higher farm incomes, a point made by David Ricardo, a noted 19th century British economist. "The price of corn is not high because a rent is paid," Ricardo wrote in 1817 in opposing Britain's corn laws, "but a rent is paid because corn is high." The extra revenue from higher crop prices is lost to rising land values as farmers bid for more land to produce even more goods to sell at higher prices.

5. a) Explain the rationale behind Ricardo's statement (in quotation marks).
   b) Which farmers will benefit from protectionism, and which will receive no benefit?

Once again America's steel industry is crying the blues and begging for protection. Worried that President Ronald Reagan's tax-reform proposal will end their chances of getting new and special tax concessions from Congress, the steel companies clamor for more import relief.

The industry still hasn't recovered from a basic error it made in 1974: to boost wages to about 175 per cent of the national average, with the expectation that consumers would foot the bill. Since then, competitors abroad have increased their advantage.

6. a) What is the meaning of the phrase (in the second last sentence) "with the expectation that consumers would foot the bill"?
   b) What would be a more realistic guess as to the industry's expectation?
   c) What would happen to wages in the steel industry if protection from imports were denied?
   d) It is often claimed that one reason the U.S. steel industry is experiencing this trouble is that it has not invested in the latest technology. What role would protection from imports have played in this decision?

The Chilliwack Inter-Church World Development Education Committee attacked Third World food production for Canada and other affluent nations.

Such production hurts both the people of the country growing the food and the B.C. farmer, Hazel Menzies of the committee said.

She said the Sonora Valley in northern Mexico produces 200 million pounds of strawberries for the North American market, which deprives Mexican peasants of the corn formerly grown on the land.

7. a) How would you rebut this argument that the Mexican peasants are hurt by the switch from corn to strawberries?
   b) Who loses and who gains in B.C.?

That's where B.C.'s problems come in. The province's economy is resource-based while Central Canada's is dependent on manufacturing. The B.C. employees of a resource company don't benefit from tariff protection because they don't need it, but they pay for tariffs to benefit industries in Ontario and Quebec.	**8.** a) What does this item suggest about Canada's efficiency in resources and manufacturing compared to other countries? b) What does this argument suggest must be done when calculating "equalization payments"? (transfers from rich to poor provinces mandated by the federal government).
The brief calls upon the government to "tackle the structural problems of phasing out those parts of the textile and clothing industry which are no longer competitive with imports, while replacing them with those industries that will be."	**9.** a) How would the free market accomplish this task? b) Why might one want the government rather than the free market to do it?
He also noted that much of the current rosy profit picture is due to the exchange rate, and should not be used as a long-term judgment on the health of the industry.	**10.** How could the "rosy profits picture" be due to the exchange rate? Explain. (Note: The writer is talking about the export industry.)
Favorable exchange rates have enabled Canada to increase lumber exports to the U.S. at the expense of a shrinking market share for U.S. forest companies.	**11.** Explain this clipping by using a supply-and-demand diagram for lumber in the U.S.
The drop in the Canadian dollar from above par in the mid-1970s to below 70 cents (U.S.) in the 1980s has helped keep Canadian companies competitive on international markets, offsetting lower productivity, said Charles Barrett, vice-president of research for the Conference Board of Canada.	**12.** Explain how the drop in the Canadian dollar helps keep Canadian companies competitive.
Profit was $15 million for the quarter, up almost 100 per cent over the first quarter of 1987, said Walsh, adding that the performance had been affected by the rise of the Canadian dollar against the U.S. dollar.	**13.** Would profits have been larger or smaller in the absence of a rise in the Canadian dollar? Explain.

Fraser Valley raspberry growers Monday were divided on the ruling handed down by the U.S. Commerce Department, which upheld a decision that Canadian raspberries are being sold at less than fair market value in U.S. markets and that the dumped berries should be subjected to an anti-dumping duty.

14. a) What do you think U.S. producers mean by "fair market value"?
    b) When could a country be considered to be a "dumper"?
    c) How will an anti-dumping duty solve the problem?

Penner added he was relieved the hearing also showed Canadian growers were not producing berries at a cost lower than production, despite charges by U.S. producers that Canadian growers can afford to do so with losses covered by farm income insurance programs.

15. The farm insurance program is intended to cover the difference between the market price and the cost of production. If price is at or above the pre-determined cost of production then no payment is made to producers. Explain how this program could lead to Canadians dumping raspberries in the U.S.

"In a free-trade environment, marketing boards could well become an albatross around the food industry's neck."

16. Explain how marketing boards could become an albatross.

Workers who suffer under free trade are like flowers that die "so that others flowers can grow," David Culver, chairman of Alcan Aluminium Ltd., said yesterday.

"You can't have growth in the garden without some deaths," Culver, a fervent free-trade booster, told reporters.

17. Which flowers will die and which will grow?

I-10 Imperfect Competition

In perfect competition each seller faces a perfectly elastic demand curve. Imperfect competition is characterized by sellers facing downward-sloping demand curves, either because they are monopolists and face the industry demand curve, or because their competition produces a similar but not precisely identical (at least in the eyes of the consumer) product. In the latter case economists talk of an oligopoly when the number of firms is quite small, and monopolistic competition when the number of firms is large (and all are about the same size).

Oligopolists play a cat-and-mouse game with their fellow oligopolists. Because there are only a few firms, the actions of any one firm have a significant effect on the profits of the other firms. Any action by one firm causes other firms to change their behavior accordingly, affecting everyone's profitability. Because of this, firms try to second-guess their competition, incorporating in their decision-making procedures a guess at how their rivals will react. Since there is a wide variety of different kinds of guesses an oligopolist could make about his rival's reactions, there are several different "theories" of oligopoly behavior, no one of which has proven to be "best."

Monopolistic competition, on the other hand, does not suffer from this dilemma. The large number of competitors, combined with freedom of entry, removes this problem and allows, as your textbook should testify, a standard theory explaining monopolistic competition. The blue-jeans market, discussed in one of the articles below, is not perfectly competitive, since jeans differ in quality and fit as well as brand name. Although this market is not characterized by a large number of sellers of roughly the same size (there are a few big companies along with smaller ones) the monopolistic competition model can be usefully applied to analyze the contents of this article.

I-10A   **Superstores**

Superstore food price war forecast

By GARY MacDONALD

Food price wars are coming.

Save-On-Foods and Safeway are waiting to kick off a competition for Lower Mainland food dollars next spring when Kelly Douglas opens its first superstore, which will be the size of a B.C. Lions playpen.

It's a war shoppers will win as grocers buy business with lower prices.

"There's going to be a major battle here when the first one opens and it will probably last 12 to 18 months," said Ian Thomas, Vancouver-based independent consultant to the retail trade.

During those 18 months, the three main discount players — Kelly Douglas's The Real Canadian Superstore, Overwaitea's Save-On-Foods, and Safeway — will butt heads as they tell Lower Mainland consumers no one beats their prices.

Says Thomas: "In the food store business, price is everything."

Executives Don Bell of Safeway and Doug Townsend of Overwaitea say superstores need sales of $1.5 million to $2 million a week to be profitable. Getting sales at those levels will mean ruthless competition.

Because they draw on a population base of about 100,000 people, superstores can devastate small independent operators whose sales drop by 10–20 per cent every time one opens near them, Thomas said.

Conventional supermarkets will be hurt the most.

"It's going to be particularly difficult for independent operators because they can't be price competitive," Eastman said.

The president of Vancouver-based Stong's grocery chain said he agrees.

"We're not going to go into a price war with them. No way," Bill Rossum said.

"We offer things the superstores can't afford," he said.

Note: Suggested answers to the questions below appear in the Appendix.

1. Using a diagram portraying monopolistic competition, show the influence on this market of the Kelly Douglas entry.

2. What does Thomas's remark that "price is everything" imply about the demand curves facing firms in this industry?

3. Show on your diagram the location of the $1.5-$2 million sales referred to in the article.

4. How would the content of the last paragraph be reflected on your diagram? Explain how it might make it possible for an independent operator to survive.

I-10B   **Blue Jeans**

'No, no, don't leave us to Levi'

By JUDY LINDSAY
Sun Business Writer

Among the fashion plates who showed up at the Vancouver hearings this week of the Textile and Clothing Board was a rangy young man whose long legs were encased in ordinary blue jeans.

He placed four pairs of blue jeans on the table in front of board chairman Gordon Bennett. The quality and fit were almost identical, he said, but one pair of jeans sells for $10 while the others go for $24.95.

He then proceeded to make a case for dropping the textile import quotas imposed by the federal government last November to protect Canadian garment-makers. Importers will be allowed to bring in only 90 per cent of their 1975 volume.

His intention, he said, was to show how some major Canadian garment companies push up the cost of their goods to Canadian buyers, and how the government with its quotas has become an accomplice.

His name is Wayne Overland, 36, and he is the major owner of six Jean Joint stores in Alberta. His philosophy, he told the board, has been to seek a higher volume of sales although it may mean smaller profit margins.

"This philosophy is apparently contrary to the policy of some of my suppliers who prefer to sell fewer units with a greater markup. They limit the supply and increase the prices.

"The Jean Joint's two best-selling brand name jeans are Levi and Howick. Both companies keep retailers on a quota system. You get only about half the units you could sell in key styles," Overland said.

He said pressure put on the manufacturers by other retailers who complained about his discount prices forced him to sell the jeans at the regular retail price.

The largest-selling jean in the world, he said, is the Levi unwashed bell bottom. Made in Hong Kong, the jean retails for $15 in the U.S. and $19.95 in Canada. The import duty to the U.S. is 16 per cent and to Canada 22.5 per cent, a difference of 6.5 per cent, he said. (That's 6.5 *percentage points*; the Canadian duty is actually nearly 50 per cent higher than the U.S. duty. Nevertheless, said Overland, it still doesn't justify the amount of the Canadian markup.)

"Then why is the price difference more than 30 per cent? It is little wonder that Fortune Magazine lists Levi Strauss as the most profitable garment company in the world."

But Overland continues to sell Levis. "Because the public is so brainwashed by their heavy advertising they'll pay more than twice the amount for the same product because of the brand name. At the same time the success of my $10 jeans indicates many consumers recognize a better deal."

A believer in "pure free enterprise," Overland said he does not advocate that the government investigate the way these companies operate, but rather it should allow small companies to compete and thus promote an "honest" market.

"Companies like mine that can offer the consumer high quality goods at low prices will do more to prevent unfair pricing than any laws.

"Levi has huge quotas for 1975 "on which they make exorbitant profits," contended Overland. "If there is a quota there will be no new importers like myself to force the prices down. This plays right into Levi's hands."

---

1. Draw a diagram representing a monopolistic competitor and use this diagram to explain the statement in the sixth paragraph.

2. What statement in the article indicates that Levis are strongly differentiated from other jeans? What impact does a more strongly differentiated product have on your diagram in question 1?

3. How does Overland's move to arrange his own supply of jeans from Hong Kong fit into the textbook theory of monopolistic competition?

4. What effect does the government's imposition of import quotas have on the adjustment mechanism described in Question 3? Does this support the accusation in the fourth paragraph that the government is an "accomplice"?

5. Interpret the final paragraph in terms of the theory of monopolistic competition.

6. Why would the price of Levis be so much higher in Canada than in the United States? (Hint: Consider Canada and the United States as separate markets with either different demand curves or different degrees of competition.)

Short Clippings

Note: Suggested answers for the odd-numbered questions below appear in the Appendix.

"Nobody is a price leader," said Greg Liddy, forest products analysts at Merrill Lynch Canada Inc. of Toronto. He noted that Canada's share of the U.S. newsprint market is declining—it has slipped to about 55 per cent from more than 80 per cent in the 1950s—reducing Canadian newsprint makers' pricing flexibility.

1. Explain what has happened over the last 30 years to Canadian newsprint firms that has reduced their price flexibility.

The crux of Titan's argument for changing the tariff was that the three eastern steel companies it had frequently asked to quote on its rod needs—Stelco Inc. of Toronto and Ivaco Inc. and Sidbec-Dosco Inc., both of Montreal—had shown only intermittent and inconsistent interest in supplying the semi-finished product.

"When they bothered to respond at all," it said, "they are at unrealistic prices as compared to world markets."

2. a) What is Titan implying about the Canadian steel industry?
   b) Why does it conclude that the tariff should be lowered?

Wayne Hekle, general manager of Canada Moving and Storage, said moving companies are battling to get as large a share as possible of the recent increase in moving business.

"The real competition is in what company is most innovative because we all have the same truck, equipment, manpower and operating costs," said Hekle.

Moving company rates, based on distance and the weight of the goods, are generally pretty competitive throughout the industry. So, where companies try to compete is in the amount of service they provide.

3. How does competition in this article differ from that described in your textbook? Explain this difference by comparing the textbook cost and revenue curves to those inherent here.

One study of the U.S. tuna fishing industry examined arrangements by which boat captains contract to sell tuna at a set price to a particular processor for a period of time.

While such an arrangement might appear to reduce competition in supplying tuna, thus increasing costs, the study concluded that it had the opposite effect.

4. Explain the rationale behind this clipping.

He said the consumer will suffer if a new carrier flies the route because increased competition will eventually cause fares to rise.

"Their fares will go down for a few months but they will soon realize that they won't make any money in the long term and fares will go back up higher than before," Lachman said.

5. a) What is the logic behind the argument that fares will ultimately rise?
   b) What is more likely to happen?

The four automakers here have a glut of about 1,700,000 unsold cars.

But when industry executives are confronted with the price question they say only that they are carefully weighing the balance between prices and production costs. At this point they say that they are going to hold prices where they are.

6. Explain how this action is rational, depending on the interest rate and expected future demand for autos.

They are the only suppliers of the meters in Canada, and have come under fire from Penticton council for selling their goods at the same price. Penticton Alderman Ron Biggs decided to ask federal Consumer and Corporate Affairs Minister Michel Cote to conduct an inquiry into possible price-fixing after Penticton received three identical bids of $28,479.12 to supply meters to that city.

7. What outcome would the textbook perfect-competition model predict in this instance? Explain.

What killed Massey Combines was declining market share, a consequence of increasing competition, in a market that was itself shrinking.

8. Explain this clipping by referring to the cost and revenue curves of a firm in imperfect competition.

During a five-year depression in the diamond business until 1986, De Beers held back diamonds from the market and substantially increased its stockpile of gems, a tactic that paid off handsomely as demand slowly recovered.

"The very essence of De Beers' operations is to keep demand and supply in equilibrium," said Neville Huxham, the company's spokesman in Johannesburg.

9. a) From the content of this clipping would you guess the price elasticity of diamond demand to be high or low? Explain your logic.
   b) What is De Beers' definition of keeping "demand and supply in equilibrium" in comparison to what is meant by this phrase in economics textbooks?

It was Adam Smith, the patron saint of doctrinaire market theorists himself, who pointed out how people in the same trade, as he put it, will consort to fix prices and overcharge the public. Market competition by itself can't be counted on to protect consumers against oligopolistic price-fixing and gouging of the public. A publicly owned participant in the market, on the other hand, with a mandate to serve the public and with no reason to squeeze excess profits out of consumers, will keep prices in check.

10.
a) Explain how and why this failure of the free market system comes about.
b) What is the usual argument countering the argument presented in the last sentence? Hint: They also have no reason to do something else.

After three years the commission concluded that any further government intervention in the economy would be inappropriate and unnecessary, provided a sound competition policy was in place. "A strong and vigorously enforced competition law is necessary to prevent dominant firms from entrenching a monopoly or quasi-monopoly position or exploiting tariff protection, to provide a check on abuses of market power, and to increase the likelihood of entry and competition from small- and medium-sized firms."

11.
a) What is meant by "abuses of market power"?
b) Explain how competition provides a check on abuses of market power.

At its latest meeting OPEC managed to achieve an unusual degree of unanimity in setting a new scale of prices and production quotas. It will be interesting to see whether the cartel has also recovered the discipline necessary to make its plans stick this time around, or whether its members will start cheating on each other—as they have done so often in the past.

The real test, especially for the economically weaker members, will come when buyers don't come rushing to buy their products at the increased posted prices of around $18 (U.S.) a barrel.

12.
a) Use a diagram to illustrate how the oil cartel would set prices.
b) Why are production quotas also needed?
c) How would a member "cheat"?
d) How is the content of the last paragraph reflected on your diagram?

I-11 Regulated Monopolies

Textbook theory of monopoly behavior shows that too little is produced at too high a price, with excessive profits, relative to the quantity, price, and profits associated with a perfectly competitive industry. Because of this result, governments often regulate monopolies to ensure that they do not exploit their monopoly position. Such regulation usually takes the form of quantity regulation (forcing railroads to run certain routes, for example), price regulation (controlling hydro prices, for example), or both.

Price regulation is tricky, because it is difficult to find the "correct" price. There are two reasons for this. First, it is difficult to get the firms in question to provide the proper cost figures, leading to endless quarrels over what the proper figures really are, as witnessed in one of the articles below. Second, it is difficult to convince governments of the relative economic merit to society of different pricing rules. For example, the government could force the firm to: (a) price so as to have marginal cost equal to marginal revenue; (b) price at minimum average cost; (c) set price equal to average cost; or (d) set price equal to marginal cost. Which of these alternatives would you choose were you the government?

Rule (a) is what the monopoly would do, if it were unregulated, to maximize profits. The theory of monopoly has already told us that this creates too high a price and not enough output. Rule (b), although it ensures production at least cost, disregards whether or not that output is wanted. Why produce 5 million frisbees of which 4 million are unwanted, just because 5 million happens to be the output at which average cost is least? Rule (c) is the rule usually adopted by government regulating agencies—it provides the good or service at a break-even price (here "break-even" includes a reasonable return on profit to the firm). Rule (d) is the rule usually advocated by economists. Although this rule does not guarantee a "break-even" operation (implying subsidies or extra profits taxes by the government) it does create the "right" quantity; rule (c) does not guarantee the "right" quantity.

What is the "right" quantity? The right quantity is that quantity for which the cost of producing the last unit equals the benefit provided by the last unit. If marginal cost is below marginal benefit, society's welfare can be increased by producing more of the good or service in question; if marginal cost exceeds marginal benefit, society will gain by reducing output. Since the price of a good measures its benefit (i.e., if the benefit were lower than the price, consumers would spend on other goods, reducing the demand and ultimately the price of the good in question), price should be set equal to marginal cost.

If this creates an excess profit for the firm, the government may simply tax it away with a lump-sum tax. If it creates a loss for the firm, however, the government is faced with the less acceptable alternative of providing a lump-sum subsidy. Why should providing a subsidy be better than making the firm price at a break-even level? If the firm priced at the break-even level there would be a higher price and a lower quantity; marginal cost would be less than marginal benefit—too few of society's resources would be allocated to this industry. If price is set at marginal cost, and a subsidy is provided, resources will be taken from a wide variety of industries in which marginal cost equals marginal benefit, and applied to this industry, in which marginal cost is less than marginal benefit. The result is a net gain to society.

All of this assumes an existing stock of capital (such as railroads and locomotives) for which the fixed costs would be incurred in any event. If production of the "right" output involves an increase in the capital stock, consideration of total rather than marginal cost must be made, since adding capital stock is now in itself a marginal change (on a larger scale).

I-11A Railroads

Former Liberal minister lambasted by Tory MPs

Special to The Province

OTTAWA—Two Alberta MPs Tuesday launched a vigorous attack against Canadian Transport Commission (CTC) president and former Liberal cabinet minister Edgar Benson.

Both Jack Horner (PC-Crowfoot) and Don Mazankowski (PC-Vegreville) tackled Benson on controversial railroad aspects when he appeared before a standing parliamentary committee.

Mazankowski, continuing a long time Western fight against railroad branchline abandonment, said government inquiries were using "misleading and distorting" costing figures to show that certain branchlines were economically unviable and should be shut down.

The Vegreville MP charged that on-line and off-line operating costs were being grouped together despite the fact that if the branchline was abandoned, the off-line costs—those relating to mainline grain shipment operations—would still remain even though farmers had to haul their grain to the mainline by some other means.

Surely, insisted Mazankowski, if branch lines were going to be saved, a fair assessment of their economic viability should be done.

Benson said the CTC had made all kinds of information available to commissions investigating branchline abandonment and grain hauling.

But, while not actually disputing Mazankowski's argument, he said Ottawa had given the various independent commissions the job of evaluating the two problems and not the CTC.

Horner, front bench opposition transport critic, charged that some branchlines were in such poor condition that trains could drive over them at no more than 10 miles per hour.

Meanwhile, in Montreal, a bus company told the CTC that the subsidization of rail passenger service inhibits the development and expansion of a passenger bus industry through the creation of artificially low fares.

The railways are charging artificially low rates for their passenger services because the federal government subsidizes 80 per cent of their deficits on those services, Provincial Transport Enterprises Ltd. said in a brief in a hearing investigating the trans-continental railway system.

Back in Ottawa, a Toronto Liberal MP charged that Transport Minister Otto Lang has raised a "red herring" in statements that it would cost $1.5 billion to improve tracks to carry passenger trains at speeds of more than 100 miles an hour in the heavily-populated corridor between Quebec City and Windsor, Ont.

Note: Suggested answers to the questions below appear in the Appendix.

1. Reinterpret Mazankowski's "on-line and off-line" argument in terms of economists' jargon of fixed costs, variable costs, average costs, and marginal costs.
2. What should a fair assessment of the economic viability of a branch line involve? Relate this to basic principles concerning profit-maximizing by firms.
3. Of what relevance to the arguments reported in the article is the fact that trains could drive over some branch lines at no more than 10 mph?
4. The article notes that a bus company is complaining about the subsidization of rail passenger service. What effect does this subsidization have on the bus company's demand or supply curves?
5. Some people claim that buses use the public roads for commercial purposes free of charge, whereas the railroads must pay to maintain their tracks, and that this constitutes a subsidy to the bus companies. Comment on the relevance of this argument.
6. Consider the railway as a monopolist on a branch line, represented graphically in Fig. I-11.1.

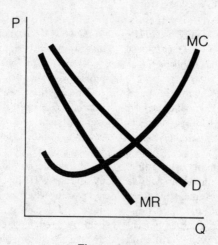

Figure 1-11.1

a) What price will the railway charge, with what resulting quantity of service provided?
b) Draw in a representative average cost curve to reflect Benson's position.
c) How can you represent on the graph the profit or loss associated with operating at the position determined in a)?
d) How does a government subsidy of 80 per cent of the railway's loss affect its price? Does this agree with the opinion of Provincial Transport Enterprises Ltd.? If not, how would this company defend itself?

I-11B Telephones

Study favors telephone competition

HULL, Que. (CP)—A government study concludes that the benefits of long-distance competition outweigh the problems and that only "a relatively small number of households would drop out of telephone service" if area rates increase slightly as a result.

The quality of service wouldn't erode much and the economic benefits to consumers and businesses would be significant under a system in which the existing long-distance monopoly is broken, a study commissioned by the federal and provincial governments says.

The communications department released the study Friday as the Canadian Radio-television and Telecommunications Commission was holding a hearing into a bid by CNCP Telecommunications for long distance competition in Ontario, Quebec and British Columbia.

Many of the economic assumptions in the study by Peat Marwick and Associates of Toronto faintly resemble those being discussed at the CRTC hearing, but the conclusion generally favors competition.

The CNCP proposal has the general support of business, but some other groups believe long-distance competition will provide the leverage for Bell Canada and the B.C. Telephone Co. Ltd. to end the historic subsidy of area rates by long-distance in what is known as rate rebalancing.

The phone companies argue that competition and other factors will force them to react by reducing long-distance rates. They say the subsidy may have to be eliminated and much higher area rates may ensue.

---

1. What does the article indicate about the elasticity of demand for area or local phone service?

2. The article indicates in paragraph five that the price for area rates is not sufficient to cover costs while the long distance rates more than cover costs. Figures I-11.2 and I-11.3 are the cost and revenue curves for the local and long-distance

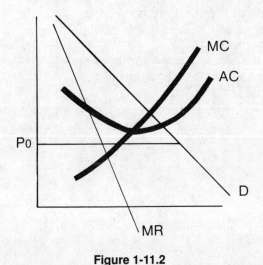

Figure 1-11.2

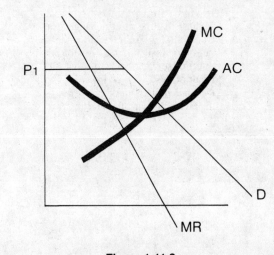

Figure 1-11.3

markets respectively. Let P_0 be the price of the area service and P_1 be the price of long-distance.
a) Identify on these diagrams the output and profits for both services.
b) As an economist can you argue for price changes in both services so that society would be better off by reallocating resources? Explain. (Hint: Think of marginal benefits and costs.)

Short Clippings

Note: Suggested answers for the odd-numbered questions below appear in the Appendix.

At its current rate of about 0.4 cents per kilowatt hour, the water charge raised $184 million in 1984. This lifts electricity prices in British Columbia close to the cost of power from B.C. Hydro's recent and future dam sites—thus contributing to economic efficiency.

1. Explain how this contributes to economic efficiency.

The relationship between electricity prices and electricity demand was long doubted by energy producers. But now electricity planners all over the world generally agree that prices do influence electricity demand, and that by matching their prices to incremental cost they can provide the right amount of electricity at the least cost.

One of the problems utilities face in setting the right prices has been determining what to do with the surplus revenue.

2. a) What name do economists give to this pricing principle?
b) How do economists define the "right" amount of electricity?
c) Why, in this example, would there be "surplus revenue"?
d) Historically, government-regulated utilities have tended to solve this surplus revenue problem by setting the "wrong" price.
 i) What is this wrong price?
 ii) Why is it "wrong"?
 iii) What name do economists give to this pricing principle?

In view of this, the Hydro studies suggest that the utility change from traditional average-based rates to rates based on expected costs of production for additional usage. The result of this would be lower bills for the prudent user, and higher ones for the lavish users in all sectors.

3. a) What terminology do economists use to refer to this pricing mechanism?
b) How would this pricing system create the result stated in the last sentence?

The entire rate structure for the pricing of electrical energy in Ontario may radically change if the directions indicated by the preliminary findings of Ontario Hydro's costing and pricing studies are confirmed and implemented.

4. Ontario Hydro is a regulated monopoly, and as such must have its rates approved by government. Suppose the monopoly can be represented by Fig. I-11.4, and assume that demand must be met (i.e., no blackouts permitted).
   a) Suppose the firm were unregulated. Use the diagram to indicate the price level, output level, and profits of the firm.
   b) Suppose the government wished to maximize the output of the firm, subject only to the constraint that the firm not suffer a loss. What quantity should it require the firm to produce? What price and profit would result?
   c) Suppose the government decided to set price so that the marginal utility of the last unit of electricity consumed were equal to the marginal cost of producing that last unit. What price, quantity, and profit (or loss) would result?

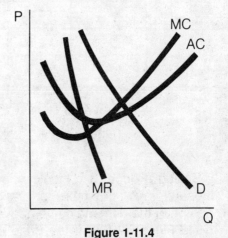

Figure 1-11.4

I-12 Price Discrimination

The term "price discrimination" refers to selling the same item to different customers at different prices. Children's rates for movies and excursion rates for airline flights are examples. Price discrimination is practised by firms because by discriminating they can raise their profits. By examining this phenomenon carefully we should be able to explain more adequately firms' pricing behavior. Although the word "discrimination" is a value-loaded term, it is not necessary that the practice of price discrimination be a bad thing. For example, although some people end up paying more for the good or service, other people, often those least able to afford it, end up paying less. It is also possible that the extra sales and revenue made possible by price discrimination could turn an unprofitable business into a profitable one, allowing society to enjoy its output rather than go without. As a last example, consider the case for charging different prices for off-peak versus peak-hour electricity, discussed in Section I-7. Here there were distinct cost savings to society resulting from the discriminatory pricing.

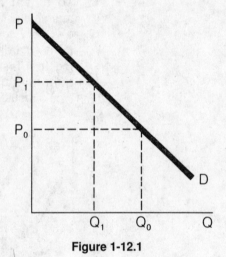

Figure 1-12.1

Selling a good at a single price to all customers is a good bargain for those willing to pay more for that good. In fact, anyone represented by that portion of the demand curve above the prevailing price falls into this category. In Figure I-12.1, if the price of airline tickets were P_1 then Q_1 people would fly; if the price were P_0, a larger number of people, Q_0, would fly. Note, though, that the Q_1 people who were willing to pay P_1 form a part of the Q_0 people. These Q_1 people now pay the airline only $Q_1 \times P_0$ dollars whereas before they paid $Q_1 \times P_1$ dollars, a difference shown by the upper left rectangle in Figure I-12.1; their consumers' surplus rises by the amount of this rectangle. Price discrimination is a means whereby the airline can prevent this rectangle from switching from producers' surplus to consumers' surplus.

Being profit maximizers, business people are always trying to find some way of transfering some or all of this consumers' surplus to themselves. If the business person is a monopolist, and can prevent resale of his output, price discrimination is a method of accomplishing this. Perfect price discrimination exists when the firm can extract from each customer the highest price he or she is willing to pay. For practical reasons this is seldom accomplished. Most price discrimination takes the form of classifying potential customers into groups (as mutually exclusive as possible) having different demand characteristics, and charging each group a different price. Children's rates for entertainment events or for transportation are examples of this.

The airlines are traditional price discriminators, using either the "excursion" versus "regular" fare or the "first-class" versus "tourist-class" fare. The two articles in this section relate to this problem of airfare discrimination. Note the important role played in both these articles by the concept of opportunity cost (the net revenue foregone by not employing a resource in its most profitable alternative use).

I-12A First-Class Travel

First class no more

First-class service simply doesn't pay, so Air Canada has finally decided to virtually eliminate first-class seats on its short-haul domestic flights—a refreshingly economical decision.

According to Air Canada president Claude Taylor, the airline will remove the first-class seats from its 52 short-to-medium range Douglas DC-9 aircraft and will reduce the number of first-class seats on its Boeing 747 jumbo jets and Lockheed Tristars. The passenger capacity of each aircraft will be increased by replacing the discarded first-class seats with economy-class.

It is no secret—and officers of Air Canada have themselves admitted—that first-class seats don't pay their own way.

They have to be subsidized through economy-class fares. Mr. Taylor explains that the purpose of the airline's program of reducing the number of first-class seats is to increase the productivity of each of its planes as much as possible—these are hard times for airlines. If that is so, why keep first-class at all, on any route? It certainly isn't a necessary service.

If Air Canada is really serious about halting its luxurious descent into the red (it lost $12 million on its operations last year), then it should either raise first-class fares to a level that truly reflects their cost or scrap them completely. Few passengers could afford the former.

Which leaves the latter.

Note: Suggested answers to the questions below appear in the Appendix.

1. The difference in cost between a first-class ticket and a tourist-class ticket is considerable, far more than the cost of the extra liquor, food and service that goes with first class. How could Air Canada deduce that "first-class service simply doesn't pay"?

2. The article asks, "If that is so, why keep first-class at all, on any route?" Answer this question.

3. Comment on the article's suggestion that "it should either raise first-class fares to a level that truly reflects their cost or scrap them completely."

I-12B   **Excursion Rates**

Money-losing Air Canada finds that if the price is right even an all-night flight to Florida can really take off

GATE 91 is at the far end of lengthy Terminal Two at Toronto's International Airport and it takes a good stiff walk to get there. But so many people are making the long hike these days that Gate 91 has become one of the brighter spots in the generally dreary state of the North American airline industry.

That's because each Friday and Saturday night, these days, several hundred economy-minded travellers crowd the departure lounge at that gate about 10 p.m., waiting to board Air Canada's new Nighthawk service to Florida.

The Nighthawk flight is getting a lot of attention from airline insiders, who have wrestled for years with how to make a dollar when your aircraft are flying half-empty much of the time.

The service uses some of Air Canada's biggest jet planes. But each Friday and Saturday night, for the past six weeks, every last seat has been filled—and all seats have been booked solidly for the remainder of the experimental program up to the end of April. And there are more than enough hopeful standby passengers to fill any booked seat that might become vacant.

Behind this phenomenon is a merchandising technique that's as old—or older—than the merchants of Baghdad. Faced with dwindling business, Air Canada's marketing men went out and cut prices—dramatically.

To fly from Toronto to Miami and back on the "Nighthawk" costs $119. To Tampa and back it's $110. That's not far off half the regular economy fares, which range between about $200 and $220, depending on the time of year and the other complex rules that determine scheduled fares. You do have to stay in Florida at least eight but no more than 60 days.

Since regular schedules of most commercial airlines are keyed to the needs of their big market—the businessman, who travels mostly morning and evening on weekdays—this pretty well dictated the time slot the Nighthawk would occupy.

It was a departure from Toronto after the Friday afternoon loads of business travellers had been cleared but before the curfew shuts down Malton flying at midnight. The times chosen were 10 p.m. to Miami and 11.15 p.m. to Tampa.

The flying time is "just under three hours," Vogan says. "The aircraft sits for about two hours in Florida then takes on another load to get back into Malton some time after operations start up again when the curfew is lifted at 7 a.m."

This full-load utilization of an aircraft for two full trips, during a period when it would otherwise be standing idle, makes the price cut possible, Vogan said, while still allowing a fair profit for Air Canada.

The same young Air Canada marketing team last year introduced excursion fares, 35 per cent below regular economy rates, at certain times on some long and short-haul routes inside Canada, and a 50 per cent youth standby discount to help fill empty seats in off-peak hours.

1. From the contents of the article, what can you deduce about the elasticity of Air Canada's demand curve versus the elasticity of the industry demand curve? Does this fit with the textbook explanation of the relationship between these two demand curves? Explain.
2. The article states that the airlines "have wrestled for years with how to make a dollar when your aircraft are flying half-empty much of the time." Comment on the viability of the following solutions: (a) cut prices; (b) cut quantity; and (c) invoke price discrimination.
3. What examples of application of the price discrimination approach are cited in the article?
4. What role does marginal-cost pricing play in this article?
5. What role does opportunity cost play in this article?
6. If you were running the Nighthawk service, would you set next year's price at the same level, a higher level, or a lower level? Why?

Short Clippings

Note: A suggested answer to Question 1, below, appears in the Appendix.

It's one thing to offer 10 pounds of ground beef in one bag at a loss leader price, because packaging, handling and other labor costs are kept to a minimum.

But it's quite another to dispense the same 10 pounds in eight to 10 smaller units, automatically increasing these costs several fold. Actually, charging a very necessary 10-cents-a-pound extra for small, sale-priced packages of ground beef does not even begin to compensate for these extra costs.

1. Can this be considered to be price discrimination? Explain in what way it is discrimination or in what way it is not.

Famous Players unreeled Wednesday a new script for ticket prices for its movie-going customers:
• It's curtains for $3 Tuesdays at Famous Players theatres across Canada.
• Canada's second-largest cinema chain will drop its Monday-to-Thursday admission price to $5 from $6.50, starting next Monday.
• Weekend and holiday admission prices remain at $6.50, while prices for children and seniors stay at $3.

2. a) What is it about the demand for movies that would cause a profit-maximizing theatre to charge more on weekends than midweek?
   b) What name do economists give to this pricing strategy?

I-13 Government Franchises

As noted in an earlier section, governments, in their role as protectors of consumers, often regulate monopolies to prevent them from taking advantage of their captive customers. This is particularly prevalent when the government itself is responsible for creating the monopoly situation, as is the case in the examples described in the following articles.

These articles relate to markets in which the government has restricted free competition by granting franchises to certain firms. Because such firms are the only ones permitted to operate, they are usually able to make above-normal profits. Monopoly profits of this nature are often collectively referred to as "economic rent"; that is, the return to a factor over and above the return necessary to prevent the factor from transferring to some other use. Such profits, and the prices and quantities associated with them, often draw the ire of newspaper editors who clamor for a change in the government taxes or regulations associated with this market.

In response, the government may establish a franchise cost—a licence fee or premise rent—as a means of taxing away the "economic rent" created by the government's own entry restrictions. Because economic rent represents an *above-normal* profit, it can be taxed away without distorting the allocation of factors in the economy.

Note: To avoid confusion, it should be noted that the term "rent" may also refer to measures established by government to tax away "economic rent" as defined above. This second meaning of the term is employed in the articles which follow.

I-13A Freeway Gas Stations

Lower rents proposed to cut gasoline prices at pumps on freeways

By PETER MOSHER

The Ontario Government may lower rental charges to oil companies operating service centres on freeways if they guarantee to pass the savings on to consumers.

Transportation Minister James Snow told the Legislature yesterday one reason gasoline prices are higher than elsewhere at 23 service centres on Highways 400 and 401 is the leasing arrangement the companies have with his ministry.

The companies' 25-year leases are based on a percentage of their gross revenues. That means they pay more as gasoline prices rise, and the Government is now taking in what Mr. Snow later called "a windfall profit out of motorists who are already hit" by high gasoline prices.

The percentages paid the Government are as high as 20 per cent Mr. Snow said, although 10 per cent was a more usual figure. The companies bid for the right to set up stations on the highways, and they have overhead expenses other stations don't have, such as the requirement they be open every day for 24 hours and provide emergency vehicles.

Outside the House, Mr. Snow said ministry officials were talking with oil company officials about an alternative. But he said he had no intention of lowering lease rentals "unless there's a guarantee the saving will be passed on to the motorist."

Note: Suggested answers for the questions below appear in the Appendix.

Suppose Fig. I-13.1 represents the demand and cost situation for a given location, exclusive of the rent paid to the government. Note that because of the location advantage, the service station can be considered a monopoly.

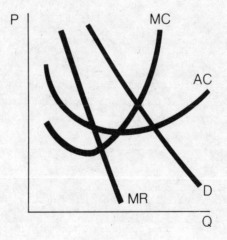

Figure I-13.1

1. Rent is usually paid as a lump sum, independent of the amount of revenue collected by the enterprise. Explain how this would be represented in the diagram and note the influence it has on price, quantity, and profits.

2. Rent could be prescribed as a percentage of profits. Explain how this would be represented on the diagram and note the influence it has on price, quantity, and profits.

3. The Ontario government has chosen to charge rent as a percentage of revenue. Note that this is equivalent to a sales tax; it can be analyzed in terms of its influence on price alone. Explain how the effects of this kind of rent can be represented on the diagram and note the influence it has on price, quantity, and profits.

4. Using your answer to question 3, explain in words how a potential service station operator would use this diagram to calculate the highest bid he would make to obtain the rights to the station.

5. The Ontario government wants a guarantee that a rent reduction would be passed along to the motorist. Would it be passed along if the government did not force them to pass it along? Explain why or why not.

I-13B Duty Free Shops

Government largesse too casual

By William Watson

TWO RECENT NEWS stories—one about duty-free shops and the other about People Express Airlines—suggest that government could be a lot smarter in the way it hands out proprietary rights to firms.

The federal and Ontario governments have been negotiating new arrangements under which Ottawa will license merchandisers to set up duty-free shops on this side of the U.S. border. This is a smart move, since it allows Canadians to pick up a share of the quite sizable duty-free business that now goes mainly to U.S. merchandisers.

What's not so smart is how the licences are going to be handed out. Though final arrangements haven't been worked out, it's likely that the procedure will be similar to that followed in Manitoba a couple of years ago.

In that province, several hundred applications were screened in a selection process that, while doubtless systematic, was also inevitably arbitrary. Once a ranking was made, the lucky winners were given their licences for free.

The best solution to the problem would be to allow anyone at all to set up a duty-free shop—the more duty-free shops the better. But trusting in competition is so alien to the Canadian nature that a true market solution to this allocation problem is the one we are certain not to adopt.

If there must be only a limited number of licensees, the next best solution to allocating them is to have an auction.

The same sort of system could be used in handing out airport landing rights. The good news recently for Canadian consumers is that People Express is setting up a Montreal-New York service. The bad news is that People's "Montreal" landing rights are at Mirabel, a large Liberal memorial situated in the scenic Laurentians.

Again the solution is obvious. Have an auction every year or two to sell off landing rights to Dorval. This way the airlines that can make the most profitable (i.e., the best) use of Dorval get to use it, while the government skims off any excess profits the airlines make by virtue of their access to Dorval.

Of course, the reason we aren't likely to move to a system in which licences are auctioned and their prices bid to a level that transfers licensees' monopoly profits to the government is simply that governments *like* having plums to hand out.

---

1. Explain why under an auction system the duty-free shops and airlines that can make the "best" use, respectively, of the duty-free licences and landing rights will end up winning the auction. Exactly what is meant by "best" here?

2. Explain how the government "skims off excess profits". What will ensure that they don't skim off too much?

3. What is the value of the "plums" referred to in the last sentence?

4. Explain why allowing anyone at all to set up a duty-free shop is even better than the auction system.

Short Clippings

Note: Suggested answers for the odd-numbered questions below appear in the Appendix.

Existing cab licences were issued by the city for $150. But they have a market value of tens of thousands of dollars, influenced by the limited number allowed by the city.

More licences should certainly be issued. But instead of charging $150, the city could auction them off to the highest bidders. Any restrictions on acquisition should be based on experience and criminal record.

1. Use a supply-and-demand diagram for cab service to explain how the market value of a cab licence is determined here.

At last the city of Vancouver has agreed to license more taxis—25 more, the first time the fleet has been increased in 38 years—and has opened tenders for the licences.

There are several very remarkable things about these tenders.

The first remarkable thing is that the highest bid for a licence is $25,000 while licences have been sold on the "black market" for up to $70,000, the price obtained for transferring ownership of any of the existing 363 cab franchises. What is apparently worth $70,000 in private trade is worth only $25,000 when the transaction is a public one and when the city is the customer.

2. Explain why this remarkable thing is not so remarkable. Use a supply-and-demand diagram to illustrate your answer.

New York's feisty Mayor Ed Koch has come up with a plan to increase the number of medallions by 1,200, or 10 per cent of the total. But the taxi interests, and the major banks which hold the mortgages on these permits, have howled in anguish and outrage, and it is not hard to see why.

3. Why?

Another common objection to a completely free market in taxi cabs is that there will be too many such vehicles on the road under such a system. If anyone is free to enter, goes the argument, everyone will.

But this objection misunderstands a basic premise of economics. Profit tends to equalize across all industries.

4. Explain how this basic law of economics answers this objection.

The absurdly irrational pricing of airport services, along with inefficiency, are to blame for the deficit. It costs exactly the same fee to land an aircraft during peak time at Toronto International Airport as does landing the same aircraft during off-peak time at Prince Rupert, B.C. The improvement in efficiency, and rational pricing under regional management reflecting cost and demand conditions, would undoubtedly reduce the federal deficit.

5. a) How would landing fees be set using "cost and demand conditions"?
   b) Why would it reduce the federal deficit?

What, then, are the lessons we can learn from the U.S. experience? First, not all regulation is the same. Strict regulation of safety is in everyone's interest; regulating prices is in the interest only of high-cost airlines. Second, many of the difficulties which have arisen since deregulation are not the result of too much market, but too little. The once-sleepy industry has developed the hyper-efficiency of a stock market, with computers fine-tuning fares thousands of times a day. But the regulators retain the gentlemen's-club habits of the past. The worst of these is the practice of awarding landing slots for free, rather than auctioning them off, thus encouraging congestion at peak landing periods, and all the attendant delays and safety worries.

6. What would determine who would get the landing slots if they were auctioned to the highest bidders?

Part II
MACROECONOMICS

Macroeconomics *Sources of Press Clippings for Articles*

II-1A	*The Financial Post*, January 9, 1989
II-1B	*The Fraser Forum*, April, 1985
II-2A	*The Financial Post*, January 12, 1987
II-2B	*The Financial Post*, September 6, 1986
II-3A	*The Vancouver Sun*, July 8, 1983
II-3B	*The Toronto Star*, June 13, 1988
II-4A	*The Toronto Star*, May 7, 1984
II-5A	*The Globe and Mail*, Toronto, January 13, 1989
II-5B	*The Financial Post*, April 12, 1986
II-6A	*The Vancouver Sun*, September 22, 1983
II-6B	*The Financial Post*, January 25, 1988
II-7A	*The Globe and Mail*, Toronto, January 6, 1981
II-7B	*The Globe and Mail*, Toronto, December 31, 1988
II-8A	*The Globe and Mail*, Toronto, February 4, 1989
II-8B	*The Vancouver Sun*, October 5, 1988
II-8C	*The Financial Post*, April 20, 1985
II-9A	*The Globe and Mail*, Toronto, January 14, 1989
II-9B	*The Financial Post*, February 8, 1988
II-10A	*The Globe and Mail*, Toronto, January 14, 1989
II-10B	*The Financial Post*, December 29, 1984
II-11A	*The Globe and Mail*, Toronto, November 16, 1988
II-11B	*The Financial Post*, February 15, 1986

II-1 Unemployment

For years the unemployment rate in Canada has been used as an index of how close to capacity the economy is operating and as a measure of welfare loss. The current character of the Canadian economy, however, is such that the traditional measure of unemployment is thought by many to be inadequate for either purpose. This is due to problems with the measurement of unemployment and changes in the character of the unemployed.

To understand the measurement problems, one must know the way in which Statistics Canada defines unemployment. This definition can be found at the back of any monthly issue of Statistics Canada's *The Labour Force* (Catalogue No. 71.001). At the heart of this definition is the requirement that to be counted as unemployed an individual not only must want a job, but also must be actively looking for one. There are reasons why the unemployment measure underestimates "true" unemployment (for example, many people pick up the odd hour of part-time work while they are looking for full-time employment, and therefore are not counted as unemployed) and reasons why the unemployment measure is an overestimate (for example, some people don't want to work but say they do in order to collect UIC benefits). The biggest roles in this measurement problem are played by discouraged and encouraged workers. As the economy moves into recession, some job seekers become so discouraged in their search that they stop looking for a job, are therefore no longer counted as unemployed, and so cause the measured unemployment rate, paradoxically, to fall. During an expansion, the opposite occurs: These people are encouraged to look for work, are therefore counted as unemployed, and so cause the unemployment rate to rise.

With regard to the character of the unemployed, the current unemployed have a much greater percentage of members of the "peripheral work force," consisting of irregular or part-time participants in the job market (see the next paragraph) rather than skilled adult workers. Using the unemployment rate as an index of capacity utilization (i.e., as an indicator of how much extra output could be produced if the economy were operating with no idle resources) is now misleading because the economy cannot easily move to this capacity without experiencing severe bottlenecks for skilled workers, with consequent inflationary pressure. With this new structure of unemployed, lowering unemployment to traditional levels would imply that the unemployment rate in certain categories (adult males, for example) would be reduced to extraordinarily low levels, too low to permit a reasonable level of frictional unemployment (workers changing jobs or retraining for new jobs), necessary for the labor force to adapt efficiently to progress and changing tastes.

Using the unemployment rate as an index of welfare loss is also misleading in the context of such a large peripheral work force, because only a fraction of those currently unemployed represent households without an adequate level of income. Almost everyone at one time or another has moved in and out of the labor force: taking a summer job, working part-time to pay for a big purchase, or quitting school to work for a year. If a larger than usual number of such people decide to enter the labor force at roughly the same time, the economy will be unable to adjust quickly enough to accommodate them, and thus they will be counted as unemployed. As a result there is great reluctance to treat high unemployment rates as seriously as they were treated in the past.

Responses to this dilemma have been either to look at different numbers for policy purposes (the job creation rate rather than the jobless rate, for example) or to create new measures of unemployment for welfare purposes (counting only those unemployed not lucky enough to have other working family members or sources of income to alleviate the consequences of their own unemployment, for example).

Another dimension of unemployment that has drawn a lot of attention is the impact of unemployment insurance on unemployment. During the years 1957–1970, average weekly benefits per unemployed amounted to about 29 per cent of average weekly earnings, but in 1971 this jumped dramatically to 41 per cent. The public began to notice a much larger number of unemployed, particularly among young people, seemingly living off unemployment insurance rather than working. Some called this "cheating," but others noted that it was simply a rational response to changes in economic incentives brought about by government policies. Newspapers delighted in running articles explaining how the unemployment insurance game "works" (how long you must work before you can collect, how much job hunting effort must be exerted to fool the government official handling your case, etc.) and then running alongside an article quoting employers who are unable to obtain employees, in spite of the high level of unemployment.

Empirical studies of this phenomenon concluded that there is in fact considerable insurance-induced unemployment, to the tune of about two percentage points. This type of "unemployment," along with frictional unemployment, structural unemployment (caused by work skills becoming obsolete, requiring retraining), and unemployment due to institutional rigidities (such as minimum wage legislation) form what is referred to as an economy's "natural rate of unemployment." The natural rate of unemployment plays a prominent role in macroeconomic theories of stagflation, discussed in a later section of this book.

II-1A Changing Unemployment

Unemployment rate's fall masks economic weakness

By Jill Vardy
Financial Post

CANADA'S UNEMPLOYMENT rate fell in December to 7.6 per cent (seasonally adjusted) from 7.8 per cent in November. But that does not mean the economy is strong.

Employment increased by 23,000 in the month, substantially less than the 66,000 jobs created in November.

Statistically, all the new jobs went to people over 25 years old, and most (20,000) went to men in that age group. Employment for those between the ages of 15 and 24 dropped by 1,000.

The unemployment rate fell so much despite the relatively small employment gain because there were slightly fewer people looking for work. The labor force—the number of people 15 years old and over working or looking for work—dropped by 1,000 in the month. There were 1.03 million people without jobs in December, Statistics Canada says.

---

1. Explain how a fall in Canada's unemployment rate could "not mean the economy is strong," as stated in the first paragraph.

2. What definition of the labor force is being used here? What role does it play in the calculation of the "unemployment rate"? How does your answer to question 1 above fit into this calculation procedure?

3. What might have caused the drop of 1,000 in the labor force, noted in the second-last sentence?

II-1B **High Wages**

Canadian Wages Are Too High

By **Herbert G. Grubel**

There are two schools of thought on the causes of Canada's high and persistent unemployment rates.

One school believes that government deficit spending is inadequately low and that monetary policies keep interest rates too high. This diagnosis is in the tradition of mainstream Keynesian economics of the post-war years. It is in disrepute in most of the world.

The second school believes that Canada's wages are too high and that profits are too low. According to this view, lower wages would enable Canadian producers to be more competitive abroad. In addition, returns to investment would increase and higher profits would generate the cash needed for making capital spending. A capital-led expansion would fuel consumer spending and ultimately permit higher wages based on increased labor productivity.

The new school's reasoning may be summarized by reference to a fact acceptable to most people. As the price of things rises, the demand falls and the supply rises, and vice versa when the price falls. Usually the price will keep changing until supply and demand are equal. Persistent excess supplies are known as unemployment in the case of labor. They are due to wages being above market-clearing equilibrium.

The school rejects the monetary-fiscal policy explanation of the persistent unemployment because it is inconsistent with recent experience. U.S. unemployment rates since 1982 have fallen by 3 points to 7 per cent while Canadian rates have fallen only from 14 to 11 per cent. Yet Canada has roughly twice the deficit, expressed as a percentage of GNP. Each has the same interest rates and inflation. The huge U.S. trade deficit and Canadian trade surplus should make Canada have a better rather than a worse experience.

The wage-profit school has further powerful, direct evidence. After 1965 Canadian real wages rose 35 per cent, and since 1976 have been unchanged. During the same period U.S. real wages rose by 10 per cent but since 1971 have fallen 5 per cent below the 1965 level. Adjustment for exchange rate changes narrows the real growth gap to about 20 per cent. Productivity increases favoring Canada during this period at most narrow this gap by one-half.

The diagnosis of Canada's unemployment problem is clear. The cure is also evident logically. The difficulties stem from finding the right medicine. The U.S. experience shows what has been successful there.

Note: Suggested answers to the questions below appear in the Appendix.

1. What in your own words is the "tradition of mainstream Keynesian economics" cited in the second paragraph?

2. Use a supply-and-demand diagram to explain the view expressed in the third and fourth paragraphs.

3. What is the clear diagnosis referred to in the first sentence of the last paragraph? What is the evident cure referred to in the following sentence?

4. What are the difficulties associated with "finding the right medicine" as noted in the last sentence?

5. Elsewhere, this author has noted that in the U.S. the rate of unionization is falling relative to that in Canada, the U.S. unemployment insurance system is less generous than that in Canada, and deregulation of industry has proceeded more rapidly in the U.S. than in Canada. How would these facts be used to help explain evidence offered in this article?

Short Clippings

Note: Suggested answers for the odd-numbered questions below appear in the Appendix.

A. Labor-Force Growth

Labor-force growth is the great unknown in forecasting unemployment rates as people's reactions are very difficult to gauge. For example, during recessions, will people not bother looking for a job because of the difficulty in finding one, or will they look even harder because of the need for additional family income when wage increases are coming in below the inflation rate?

1. Explain what impact each of these two options would have on the unemployment rate.

In any economic slowdown, the unemployment rate rises, first and foremost, because employment growth slows or actually goes into negative territory. But how high the jobless rate goes can also depend very crucially on exactly how fast the labor force decides to grow.

2. Explain how labor-force growth plays a role here.

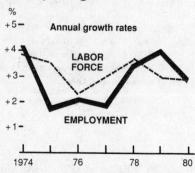

How jobs grow

3. What must have been happening to the unemployment rate: a) during 1974, b) during 1977, and c) during 1979? Explain your reasoning. (Note: The growth rate throughout a year is given by the single point corresponding to that year.)

THE UNEMPLOYMENT rate moved up to 7.8 per cent (seasonally adjusted) in May from April's 7.7 per cent, but not because of job losses. Indeed, employment rose by a big 0.6 per cent or 68,000 during the month. However, that wasn't as big as . . .

4. Complete the last sentence.

Job creation has been impressive. But the national unemployment rate this past June was higher than in pre-recession June 1981—7.7 per cent versus 7.1 per cent. Moreover, it was significantly higher in eight of the 10 provinces.	5. How could the unemployment rate be higher when so many new jobs were created?
Unfortunately, it's going to be at least the 1990s—and perhaps well into that decade—before we return to a 7.5 per cent jobless rate. The timing depends on productivity and on labor force growth.	6. Explain the influence of productivity and labor-force growth here.
The drop has been driven by the appearance of 304,000 new jobs, which pushed total employment to 11.47 million last month. Over the same period, only 204,000 people have entered the job market.	7. a) To what does the word "drop" refer? b) Of what relevance here is the information about the number of people entering the labor market?
The council, an advisory body with business and union representation, decided to rethink the concept of unemployment in late 1972. It was intent on unraveling the paradox of substantial new job creation coupled with a high and largely unyielding jobless rate.	8. How could this paradox come about?
"The 1970s," he said, "have to be seen as a period when supply shocks and structural change combined to produce a stagflation that no demand-oriented policy weapons could be expected to reach. The oil crisis and women's invasion of the labor force come readily to mind."	9. Explain how the reference to women's invasion of the labor force makes sense.

B. Participation Rate

In fairness, the strong female job growth occurred partly because so many more women were available to work. The female participation rate rose to 55.1 per cent in 1986 from 45.2 per cent a decade earlier. That translated into an extra 1.7 million women in the labor force. In contrast, the male participation rate fell during the period, to 76.7 per cent from 77.6 per cent, which meant that there were only an extra one million men in the labor force.	1. a) What is meant by the phrase "female participation rate"? b) Can we conclude from this information that the female unemployment rate must have fallen relative to the male unemployment rate? Why or why not?

II-1 Unemployment 131

Indeed they look for pretty sluggish job growth during the rest of the year. That's likely to keep the unemployment rate in the 10.5 per cent–10.75 per cent range, particularly as the recent increases in employment will probably encourage more people to join the labor force.

2. Explain the rationale behind the last half of the second sentence. What terminology do economists usually employ to refer to this phenomenon?

The current unemployment rate has fallen to just below 11 per cent. The reduction in the unemployment rate which is projected to take place as a result of the budget proposals is expected to be moderated as participation rates continue to increase; as a result the unemployment rate is forecast to decline to about 10.25 per cent by the end of 1986.

3. Why would rising participation rates moderate the reduction in the unemployment rate resulting from the budget?

The 1980s look entirely different. For instance, the Finance Department expects both employment and labor-force growth to average 2.2 per cent a year in the 1980–85 period. This assumes that the participation rate (the percentage of working-age people who are working or want to work) continues to increase at a fair clip, from 1979's 63.3 per cent to 66.5 per cent in 1985, similar to the 3.6 percentage-point increase from 1973 to 1979. This is the critical assumption in the projection. If participation rates turn out to be higher, the unemployment rate will be higher and vice versa for lower participation rates.

4. a) By means of a supply-and-demand diagram, explain why a higher participation rate would increase unemployment.
b) What should it do for employment? Explain how.

Labor-force growth has, in the past, tended to reflect the economic situation. In all post-war recession years, except 1960, the participation rate fell slightly, with the result that. . . .

5. Complete this clipping by explaining how this phenomenon would affect the level of unemployment as the economy moved into recession.

C. Discouraged and Encouraged Workers

"We consider it entirely possible that an "encouraged worker" effect will set in when employment picks up, which could lead to a paradoxical rise in the unemployment rate," he said.

1. Explain how this paradox could come about.

A closer look at the numbers shows little to cheer about. 　The unemployment rate didn't go down because more jobs were created as Finance Minister Jean Chretien implied when he called it "good news" and a "good trend" in the House of Commons.	2. a) How could the unemployment rate go down if more jobs weren't created? 　b) Why might the author of this clipping feel that your answer to a) implies that there is "little to cheer about"?
For example, Canada's jobless rate fell to 7.5 per cent last month, the lowest number in 2 1/2 years. But if you think it's a sign that the economy is suddenly moving up, look again.	3. Why wouldn't a fall in the unemployment rate be a "sign that the economy is suddenly moving up"?
The working-age population is expected to increase by about 2.6 per cent this year. But the labor force could well increase by a much greater amount, as the so-called "additional worker" effect outweighs the "discouraged worker" effect, and sends the participation rate up.	4. What are the additional worker and discouraged worker effects? How could they affect the unemployment rate?
Thus, the marginally-attached worker prevents the jobless rate from dipping in good times, and accelerates it in bad times.	5. What argument could generate this conclusion?

D. Unemployment Insurance

Unemployment insurance may increase the incidence and length of joblessness and has certain inflationary attributes, but these drawbacks are outweighed by the positive social factors, according to a new study published by the Brookings Institution.	1. How does unemployment insurance 　a) increase the incidence of joblessness; 　b) increase the length of joblessness; 　c) have inflationary attributes?
"The similar time pattern in withdrawals from the labor force probably indicates that some workers who would have left the labor force held out until their benefits were exhausted."	2. Explain what impact this behavior has on the measured level of unemployment.

Marston said unemployment insurance can cause insured workers to remain unemployed longer than they otherwise might because they can afford to be more selective about the jobs they will accept. Also, it allows a person with minimal job prospects to collect a cheque although technically he should drop out of the labor market until conditions improve.

3. What impact would these phenomena have on the unemployment rate?

There are some subsidy features to the insurance, primarily for jobs that are unattractive or that offer seasonal or unstable employment. Without this subsidy, paid by the employer and government, Marston said many unskilled jobs would go begging.

4. a) Explain how unemployment insurance subsidizes certain kinds of jobs.
   b) What do you think would happen to the wages paid in these jobs if unemployment insurance ceased to exist?

Mr. Reid is critical of the flat rate contribution schedule for unemployment insurance, which he says acts as a subsidy from companies with more stable employment levels to those which frequently lay off workers. "This subsidy encourages firms to rely more heavily on layoffs of workers than other methods of adjusting to fluctuations in the demand for their output." He suggests that the excessive and inefficient bias toward layoffs in the Canadian system could be eliminated by adopting a system of experience rating—a company's contributions would be related to the amount of benefits drawn by its employees in the past.

5. a) Does the mere existence of unemployment insurance subsidize seasonal industries? Explain why or why not. Comment on whether or not wages in seasonal industries would be higher or lower in the absence of unemployment insurance. (Note: Flat rate contribution is one in which everyone contributes at the same rate per employee.)
   b) Would adoption of experience rating raise, lower, or leave unchanged wages in seasonal industries? Explain your reasoning.

"The rationale is simple. Unemployment insurance is supposed to protect against unexpected income loss. We don't give low-cost life insurance to pyromaniacs, or auto insurance to reckless drivers. Similarly, we should pass the high cost of income protection on to workers who negotiate in a manner that increases the likelihood of future income loss. The insurance industry would call this 'experience-rating.'"

6. Explain in your own words what this suggestion amounts to.

E. Miscellanea

In a typical recession, the unemployment rate moves up sharply, although it can take a few months for this to happen since employers are loath to lay people off until they're sure sales are really down. Once the recession is over, the rate usually moves down again, although again there is a lag as employers don't start hiring until they're sure of the recovery.

1.
a) Why would employers be slow to lay people off? (This is called "hoarding" labor.)
b) Use the concept of labor hoarding to offer an alternate explanation for the phenomenon cited in the last sentence.

As this argument has it, a major easing in the rate of inflation has been hamstrung by a significant decline in productivity over the past year. When productivity "turns favorable, that is when unemployment is high but falling, the rate of inflation could decelerate markedly for something like a year."

2.
a) Why should productivity turn favorable when unemployment is high but falling?
b) Why should this cause inflation to decelerate?

Yet despite the changed composition of the jobless ranks, today's high unemployment level is portrayed nonetheless by some as a disaster of Great Depression proportions.

3. Of what relevance here is the changed composition of the unemployed?

According to the Finance department, output is projected to revive at 4.2 per cent a year. Given a 1–1.5 per cent yearly increase in productivity that one may expect over a period of gradual recovery, this translates into an increase in employment starting in 1984 of about 3.5 per cent. Since the *trend* rate of increase in the labor force is just less than 2 per cent per year, that would seem to produce a healthy decline in the unemployment rate of 1.5 percentage points per year.

Unfortunately, matters are not so simple because, as is by now well understood, the normal "official" unemployment rate does not include all the sources of slack in the labor market.

4.
a) Explain the logic behind the calculation of a 1.5 per cent annual decrease in the unemployment rate. (Hint: The arithmetic in this clipping is incorrect.)
b) Give an example of one of the "sources of slack" referred to in the second paragraph, and explain how it impacts upon the 1.5 per cent figure.

The proportion of young men who have jobs has jumped, in the past two months, to more than 60 per cent.

The employment rate—or employment ratio, as it is sometimes called—is a relatively new statistic, little used so far to chart the course of the economy or forecast its future.

But Statistics Canada has been publishing employment rates for various categories of Canadian workers for a year now and some of them—especially the rate for men aged 15 to 24—show promise as sensitive indicators of shifts in the economy.

5. a) Define the employment rate.
   b) Why would the employment rate for men aged 15 to 24 be a good indicator of cyclical change in the economy?
   c) Why would the employment rate be better than the unemployment rate in this regard? Isn't one just 100 per cent minus the other?
   d) From the information in this clipping, at what stage in the cycle is the economy at present? Explain.

Since 1960, when the average work week in Canadian manufacturing reached 40 hours, it has gone up with economic expansion and declined in recession.

6. Explain why this phenomenon occurs. (Note: Average figures refer only to those working.)

The unemployment problem must be seen as a matter not of sadistic central bankers or fate, but of *insiders* vs. *outsiders*. That is, a surplus of unemployed labor—the outsiders—is most often the result of privileges afforded the insiders who have jobs.

7. a) What is the meaning of the reference to "sadistic" central bankers?
   b) What are the "privileges afforded the insiders"?

II-2 Consumption And Aggregate Demand

The Keynesian approach to the determination of national income focuses on the level of aggregate demand for goods and services, of which consumption demand is by far the largest ingredient. The Keynesian multiplier process, explaining how an increase in demand can lead eventually to an even greater increase in income, operates primarily through the consumption function: A stimulus to demand increases income which in turn increases consumption, continuing the expansion.

Keynes introduced the concept of the consumption function in his famous book, *The General Theory,* in 1936, and defended it on the basis of a fundamental psychological law. Although the simple Keynesian version of the consumption function (in which consumption is a function of income) did fairly well in predicting consumption in the late 1930s, it failed disastrously in predicting post-World War II consumption. During the war, rationing and patriotism led people generally to abstain from consumption and accumulate their savings. At the end of the war, the populace had accumulated considerable wealth and was ready to buy consumption goods in excess of what their current income would have permitted; the Keynesian consumption function should have included the wealth level in addition to income to predict consumption.

Because of this spectacular failure in predicting consumption, economists devoted considerable energy to the development of theories of consumption behavior. In newspapers these theories tend to appear in the guise of reports on saving behavior.

The textbook 45° line diagram is the vehicle usually employed to illustrate the role played by aggregate demand for goods and services. Many textbook expositions also utilize the aggregate supply-aggregate demand diagram, which allows the supply side of the economy to play a role in the determination of national income. In answering the questions in this section, use the diagram with which you are more comfortable.

II-2A Uncertain Saving

Experts divided on uncertain savings outlook

THE ECONOMY

BY CATHERINE HARRIS

CONSUMERS HAVE been providing a big engine of growth for the Canadian economy since the 1981–82 recession, increasing spending at a very rapid pace, despite weak wages gains and higher personal income taxes.

This has been made possible by consumers' willingness to deplete savings. The personal savings rate fell to 9.1 per cent (seasonally adjusted) in the third quarter of 1986 from a peak of 18.1 per cent in the first quarter of 1982.

The critical question now is whether a 9 per cent savings rate can be sustained. If consumers decide to rebuild savings, that will take away from personal consumption and result in lower overall economic growth.

Economists are divided on what the savings rate is likely to do. So we get the usual wide range of economic forecasts.

A lot of factors can affect the savings rate—inflation, interest rates, unemployment, the tax system and demographics.

•**Interest rates** encourage savings when they're high and discourage them when they are low. The question, of course, is what's high or low.

The sophisticated consumer looks at real interest rates. That is the gap between interest rates and inflation. If it's high, then he saves a lot; if it's low, then he saves less. Currently, real interest rates are pretty high.

However, the unsophisticated consumer tends to look at the actual level of interest rates in comparison to the recent past. And, as Steven Tanny of Woods Gordon points out, current interest rates don't look very high in comparison with the last four or five years.

Thus, interest rates could be having the effect of encouraging some consumers to save, but at the same time discouraging others.

•**The misery index** (the unemployment rate plus inflation) can also affect savings, says Leo de Bever of Crown Life Insurance Co.

High unemployment spells job insecurity and the necessity to build up savings for use in a possible period of unemployment. High inflation means things will cost more down the road, so savings have to build up, in order to ensure enough funds to maintain one's standard of living. Thus, a high misery index encourages savings while a low one tends to lead to a relatively lower savings rate.

The misery index is currently relatively low; indeed at 13.8 per cent in 1986 (a 9.6 per cent unemployment rate and 4.2 per cent CPI rise) it's at its lowest level since 1973. This factor should, therefore, be encouraging spending rather than savings.

•**The tax system** can have a powerful effect on savings. And certainly the Canadian tax system encourages savings—through RRSPs, the dividend tax credit, and the $1,000 interest and dividend deduction. And with RRSPs limits rising, the system is likely to continue to encourage high savings.

However, the tax system will only encourage higher savings if people can afford to use all the deductions. And given higher personal taxes and modest wage gains (coming in lower than the rise in the CPI in all but two years during the past nine years), consumers may be losing their ability to use up all the potential deductions.

•**Demographics** should be encouraging savings, since a growing proportion of people are moving into the high savings age-bracket. Specifically, the baby-boomers are in their prime earning years and, thus, are able to put aside funds to pay off mortgages or to provide for retirement.

Note: Suggested answers to the questions below appear in the Appendix.

1. Explain how consumers can provide an "engine of growth" for the economy. (first paragraph)

2. From the numbers cited, which can you calculate—the average propensity to consume or the marginal propensity to consume? What is its current value?

3. Explain in your own words, using a numerical example, how interest rates could be encouraging some consumers to save, but discouraging others, as noted in the ninth paragraph.

4. Consider the argument presented in the second paragraph of the section dealing with the tax system. Does this suggest that the marginal propensity to consume is larger or smaller at higher income levels? Explain.

5. At several points in this article, reference is made to a desire to build up savings. What impact does this have on consumption demand and on the level of national income? Explain.

II-2B GDP Rise

Danger signs lurk behind GDP rise

REAL gross domestic product rose 0.8 per cent (seasonally adjusted) in the second quarter, on the strength of a 1 per cent increase in consumer spending.

That's quite encouraging, but we're not out of the woods yet.

Business capital spending dropped a big 4.6 per cent in the quarter and may well decline again in the third quarter as the impact of lower energy prices continues to make itself felt on energy investment.

Exports dropped, too, in the quarter, by 1.8 per cent. That was overshadowed by a 4.4 per cent decline in imports but we won't feel secure in our growth prospects until exports pick up again.

Also discouraging was continued inventory accumulation, almost equal to the first quarter's big increase. That means we're likely to see lower production down the road as those inventories are used up.

And finally, the rise in consumer spending was partially fueled by a big drop in the savings rate, to 11.2 per cent from 12.5 per cent in the first quarter. We could easily see that partially reversed in the third quarter since we haven't had a savings rate that low for a sustained period since 1977. And that would mean a lower gain in consumer spending.

---

1. Show on a 45° line diagram, or on an aggregate demand-aggregate supply diagram, how to portray:

    a) a 1 per cent increase in consumer spending (first paragraph);
    b) a 4.6 per cent drop in business capital spending (third paragraph);
    c) a 1.8 per cent drop in exports (fourth paragraph);
    d) a 4.4 per cent decline in imports (fourth paragraph); Warning: The answer to this question is not straightforward.
    e) a continued inventory accumulation (fifth paragraph).
    f) a partial reversal of the first quarter's drop in the savings rate (last paragraph).

2. With the help of your diagram from question 1, explain how the result stated in the second sentence of the fifth paragraph will come about.

Short Clippings

Note: Suggested answers for the odd-numbered questions below appear in the Appendix.

A. Savings and the Savings Rate

THE LOWEST savings rate in almost 15 years points to slower consumer spending—and thus lower growth—ahead.

According to Statistics Canada, the savings rate tumbled by a full percentage point to 9.5 per cent in the first quarter of 1987, compared to the previous three months, as consumers used savings to help finance their purchases.

1. a) Is this reflected by a change in the MPC or the APC? Up or down?
   b) The word "savings" in the second-last line does not have the same meaning as in the first line. Explain this difference.

The federal Government will be prevented by the size of the deficit from taking any action to stimulate the economy. Real wages will continue to fall, and the only possible source of stimulus is a decrease in the personal savings rate, he said.

2. Explain how a decrease in the personal savings rate could act as a source of stimulus.

Earlier figures released by the agency also showed that the savings rate—savings as a percentage of personal disposable income—fell to 7.5 per cent in the first quarter of the year from 8.3 per cent in the final quarter of last year.

3. a) Is this reflected by a change in the MPC or the APC? Up or down?
   b) By means of a diagram, show what impact this change in the savings rate would have on the overall level of economic activity.

All this frugality is producing a consumer-generated recession. What is needed is some policy action to stimulate consumer spending to move the economy out of its current lassitude. Consumer attitude surveys indicate that the time is ripe for such a move, and the high savings rate of recent years has led many economy-watchers to predict an upturn.

4. a) What is a consumer-generated recession?
   b) How would higher consumer spending move the economy "out of its current lassitude"?
   c) Why does the high savings rate of recent years point to an upturn, as stated in the last sentence?

Savings banks hold deposits equal to almost $2,000 Cdn for every man, woman and child in the Soviet Union, which reflects how difficult it is for people to find anything worth spending their money on.

The per capita savings figure of about 1,000 rubles ($1,940 Cdn) represents half a year's salary for most Soviet workers and the hoard could generate serious inflation when more consumer goods reach the market.

5. Do these figures suggest that the MPC in the Soviet Union is approximately 0.5? Why or why not?

B. Consumption Behavior

Rising interest rates may cloud what has been the very sunny disposition of consumers, the Conference Board of Canada said yesterday.

The board's latest quarterly survey of consumer attitudes, conducted in the second quarter of the year, showed "consumer confidence is again at near-record highs."

1. What two factors, in addition to the income level, does this clipping suggest ought to be included as explanatory variables in the consumption function?

The trouble, of course, is that Canadians do not necessarily think like central bankers. Says Beigie: "I don't think the consumer spends on the basis of saying, 'If the Bank of Canada raises interest rates 50 points, then I'll cut my spending rate from 92 per cent of my income to 91 per cent.' They're saying, 'I feel good, so I'm going to spend. I don't give a damn what those clowns in Ottawa are doing.'"

2. a) What does Beigie think is the consumption demand elasticity with respect to the interest rate?
b) What does he think is the primary determinant of changes in consumption spending in the absence of income changes?

Another element not taken into account immediately after the crash was the so-called "wealth effect." It was quickly revealed that less than 20 per cent of Canadians owned stock-market investments anyway.

3. a) What is the "wealth effect"?
b) Of what relevance is the fact that "less than 20 per cent of Canadians owned stock market investments"? Note: The "crash" refers to the October 17, 1987 stock-market plunge.

The value of retail sales in November rose 0.6 per cent (seasonally adjusted) from October. That's not as high as the average monthly increase of 1 per cent for the first 10 months of 1987, but it still reflects solid growth. Of November's rise, half was an increase in volume since retail prices were up only 0.3 per cent in the month.

4. Explain how this author gets the result that "half was an increase in volume" as noted in the last sentence.

"Our biggest problem in the present tax system is that the amount of personal income tax Canadians are paying is too high.

"The government believes that personal income taxes should be reduced to leave more money in the pockets of Canadians to spend or save as we see fit. This means we will be seeking higher revenues from the corporate and sales tax systems."

5. From this information, would you guess that Canadians would "spend or save" the "more money in their pockets." Explain.

The federal government could encourage more consumer spending to boost the economy by cutting tax breaks on home ownership and retirement savings plans, says the chief economist for the Conference Board of Canada.

6. Explain the rationale behind this statement. Wouldn't cutting tax breaks reduce disposable income and thus reduce consumption spending?

C. Influence on Income

Canadians saved 17.8 per cent of their disposable income during the recession year 1982, the highest rate since the Second World War. The level of savings has fallen off since then to less than 10 per cent of disposable income.

1. a) Is this reflected by a change in the MPC or the APC? Up or down?
b) Would the higher savings rate during 1982 have eased or exacerbated the recession? Explain how.

The slack exists, of course, because people continue to save even when the private sector demand for loans ebbs away. And they tend to save more, not less, in scary times.

2. Does this activity serve as an automatic stabilizer or destabilizer? Explain.

Most economists these days are preaching that consumers had better start spending.

Economic Council of Canada chairman David Slater: "Fears of unemployment have curtailed consumer spending and starved the federal government of tax revenues needed to reduce the deficit."

3. How has a curtailment of consumer spending "starved the federal government of tax revenues"?

The fuel for Canada's relatively good economic performance last year came from a strong export sector but we are now becoming more dependent on a recovery of consumer spending as export growth slows.

4. Explain how exports and consumer spending act as "fuel".

Yet, in recent months, economists have been noticing that consumer spending, which has fueled much of the economic growth of recent years, is being replaced by business investment and export sales as the engines of economic growth.	**5.** Explain how consumption can act as an "engine of economic growth."
The only other potential bright spot observers spy on the horizon will be lower personal tax rates and higher take-home pay under the federal tax reform package, that will begin July 1, 1988.	**6.** Why would this be viewed as a "bright spot"?
"The stock-market crash of 1987 really is an amazing phenomenon," says economist Patti Croft at investment dealer Burns Fry Ltd. in Toronto. "Just six months ago every economist was looking for a recession, if not outright depression."	**7.** What do economists see as the link between a stock-market crash and a recession?
Previously, renovations and repairs to houses, as well as spending on carpets, glass, paint and wallpaper, were counted as consumer spending. Now they are counted as residential investment and, as such, they become a form of savings by householders. As a result, 25 years of figures for consumer spending have been revised downward, while the figures for residential spending—and savings—have been shifted up.	**8.** Do these revisions imply a need for revisions in the national income figures? If so, in what direction? If not, why not?

II-3 Investment and the Inventory Cycle

Investment plays a key role in the economy. As a component of aggregate demand, it is an important element in the maintenance of full employment, and as a contributor to the capital stock it increases the efficiency with which the economy operates and augments longer-run aggregate supply capacity.

Because of these active roles played by investment in the operation of the economy, economists have expended considerable energy in the search for the determinants of investment. Unfortunately, their search has not been a successful one; for there appears to be a large number of variables affecting investment behavior, each operating with different intensity under different circumstances. For example, when excess capacity exists in the economy, encouraging investment usually meets with little success.

The determinants of investment most often described in textbooks are the interest rate, the level of income, changes in the level of income, profits (or retained earnings), and special government tax allowances. All of these appear in newspaper accounts; some are referred to in the shorter clippings below. One particular dimension of investment demand, inventory changes, appears frequently in newspapers. Changes in inventory levels are a key ingredient of investment demand. Most students in introductory economics courses will identify their importance in one of three roles. First, inventories make the "accounting identity" work. In each year total income (what is produced) equals consumption plus investment plus government spending (what is demanded), according to the accounting identity. But what if the total produced isn't all demanded? Thanks to inventory changes, this cannot happen, since they are *defined* as being part of business investment. If total demand falls short of the total produced, inventories will rise (a rise in investment demand), increasing "total demand," making it equal the total produced.

Second, rises and falls in inventories act as messages to producers, telling them whether they are producing too much (or pricing too high) or producing too little (or pricing too low). Thus the behavior of inventories is a critical ingredient in any explanation of why the economy is doing what it is doing when it is moving from one equilibrium position to another.

Third, inventory levels, the changes therein, are useful for forecasting. Suppose inventories are at a very high level. Firms will cut back on production (meeting demand out of inventory), to bring inventories back to their desired level; we can therefore anticipate a fall in income. As another example, suppose inventories are falling. This means that aggregate demand for goods exceeds the quantity being produced; producers should soon get this message and increase production, so that we can anticipate a rise in income in the near future.

In more advanced economic analyses, inventory behavior plays an important role in the dynamics of the economy, particularly with respect to business cycles. This stems not from their size (inventory changes usually comprise less than five per cent of total investment demand) but from their volatility.

II-3A **Recovery Stronger**

Recovery stronger, inflation held down

OTTAWA (CP)—Statistics Canada continued to take an upbeat tone on the economic front Thursday, saying evidence suggests the recovery has built sufficient momentum to be self-sustaining for the immediate future.

The report, based on an analysis of data available to May 13, also said the recovery appears to have had no adverse impact so far on inflation, which by May had slowed to an annual rate of 5.4 per cent from 11.9 per cent a year before.

"An increase in productivity that is typical during the early stage of expansion appears to have enabled firms to improve their profit margins without raising prices significantly," the report said.

However, the federal agency warned the path to long-term economic recovery is not without hurdles.

"The major concerns for the longevity of the recovery appear to remain the continued high interest rates, a possible faltering of the global recovery, or a resurgence of inflation."

The report attributed the strength of the recovery during the first three months of this year to increased consumer spending and residential construction, as well as a reduced rate of inventory liquidation.

Prospects for continued consumer demand appear good because consumer confidence and the expectations of manufacturers are up, it said.

Note: Suggested answers to the questions below appear in the Appendix.

1. What mechanism is implicitly referred to in claiming that the recovery will be self-sustaining? (first paragraph)
2. Why is an increase in productivity typical during the early stage of expansion? (third paragraph)
3. Why would "a reduced rate of inventory liquidation" be responsible for the strength of the recovery, as claimed in the second-last paragraph?
4. Explain why continued high interest rates and a faltering global recovery could prevent the recovery from being a lasting one. (fifth paragraph)
5. How does consumer confidence affect the recovery? (last paragraph)

II-3B Investment Surge

Business investment surge 'out of phase' by Arthur Donner

Canada is experiencing a major wave of new business investment at the tail end of a long business cycle expansion. This latest surge of business investment will make it very difficult for the Canadian economy to slip into recession in 1989, even if the U.S. economy turns down. However, the added industrial capacity will also result in much weaker investment spending in the early 1990s, which may restrain Canada's economic progress at that time.

The "out of phase" character of business investment spending relative to the rest of the economy is quite typical for Canada.

Investment spending represents a vote of corporate confidence in the way the economy operates. Because of the drawn-out nature of the investment planning process, new investment decisions are never taken lightly. Nevertheless it is almost remarkable the extent to which a herd instinct operates with respect to the timing of new investments. Most businessmen are either bulls or bears with regard to capital investment—at about the same time.

Business investment usually represents only 10 to 15 per cent of Canada's GNP, yet it is always regarded as one of the key indicators of the health of the economy. Business investment is also highly volatile. Generally speaking, most private investments are regarded as positive for the economy; however, timing, circumstances and time lags all play important roles in assessing the impact of investment spending on the economy.

For example, new large scale investment projects can be inflationary if they are piled on top of an overheating economy. Yet investment spending ultimately increases industrial capacity, which is usually regarded as anti-inflationary.

Investment also correlates closely with job creation, though in the Canadian economy the import content of machinery and equipment investment is so high that the country's foreign trade balance often worsens during investment booms.

The last major capital spending wave in Canada occurred between 1979 and 1981. That investment wave was triggered by the second doubling of world oil prices by the members of the Organization of Petroleum Exporting Countries.

Indeed capital investment spending remained robust in 1981 even as the Canadian economy was plunging into its worst recession in a half a century. Business investment spending in 1981 initially moderated Canada's slide into recession, but as the recession deepened in 1982 business investment began to contract as well, which made matters even worse for the Canadian economy.

---

1. Explain why "This latest surge of business investment will make it very difficult for the Canadian economy to slip into recession" as stated in the first paragraph.

2. How could a Canadian recession be created "if the U.S. economy turns down"? (first paragraph)

3. Explain why investment spending will be weaker in the early 1990s, as noted in the first paragraph.

4. Explain why "Investment spending represents a vote of corporate confidence in the way the economy operates," as stated in the third paragraph.

5. Explain how investment spending can be viewed as both inflationary and anti-inflationary, as claimed in the fifth paragraph.

6. What does the import content of machinery and equipment have to do with the correlation between investment and job creation? (sixth paragraph).

7. Explain how business investment could moderate Canada's slide into recession in 1981 but make things worse in 1982. (last paragraph)

Short Clippings

Note: Suggested answers for the odd-numbered questions below appear in the Appendix.

A. Interpreting Inventory Changes

Manufacturers ended last year on a strong note with both shipments and orders rising to record levels in December, Statistics Canada said today.

At the same time there was an ominous buildup of inventories, which also rose to a record level.

1. Why is the buildup in inventories described as ominous?

Although initially the news looked good, with growth of 4.2 per cent in the final three months of 1987, closer examination showed a sharp drop in consumer spending and a large expansion in inventories.

2. Explain why this closer examination makes the news not look good.

When the news of Wednesday's numbers on the gross national product was made public, stocks and bonds immediately rose and the U.S. dollar strengthened.

Then, when analysis of the numbers came in, the markets went into reverse.

The reason was that the greater part of the improvement in the quarter—$33.7-billion (U.S.) out of a total GNP advance of $39.2-billion—came from additions to business inventories.

3. Explain why the markets would go into reverse upon learning about the additions to inventory.

Because consumer buying has held up surprisingly well during this recession, key measures of the burden of inventories—the so-called inventory-to-sales ratios—have improved markedly.

4. Why would the inventory-to-sales ratio be a measure of the "burden" of inventories?

Pedderson said the news on U.S. growth and inflation "is cause for some happiness." But he noted that the latest gain was mainly due to an increase of inventories, especially unsold cars. This will translate into slower growth in the second quarter, he said.

5. Explain why "slower growth in the second quarter" will occur.

The inventory buildup, the major source of strength in the first quarter, will not be repeated and will actually be a source of weakness as production is reduced to work down unwanted stockpiles in the face of slumping sales, analysts said.	6. Explain in your own words how an inventory build-up can be *both* a source of strength *and* a source of weakness.
The extent of the downturn surprised many analysts. According to Statistics Canada, the real gross national product fell at an annual rate of 4 per cent . . . If an apparently unplanned buildup of inventories is taken into account, the rate of decline is estimated at 7.6 percent."	7. What is the logic for taking into account the unplanned build-up of inventories?
The GNP grew at a much faster 4.8 per cent rate in the October-December quarter, but analysts said that growth masked some dangerous imbalances that were not present in the first quarter report. Almost all of the fourth quarter increase in GNP wound up as unsold inventory sitting on shelves.	8. If it doesn't get sold, how could it get counted into GNP?
In the first two quarters of the year, consumer spending was somewhat sluggish but there was a strong buildup of inventories by business which boosted economic output, Saba explained.	9. Does it matter if output is boosted by strong consumer spending or by strong inventory buildup? Why or why not?
Changes in inventories go a long way, for instance, toward explaining why first-quarter real GNP was ahead of a year earlier by a meagre 1.7 per cent.	10. What must have been happening to inventories during the first quarter?

B. Forecasting with Inventories

The main factor behind those projected rises is a stronger economy. He says increases in U.S. consumer spending, accompanied by reductions in inventories, have set the "stage for an awfully strong second half."	1. Why would a) increases in consumer spending, and b) reductions in inventories foretell a stronger economy?
Michael Manford told a conference on the outlook for fixed income markets that Canada is headed for a big inventory liquidation that will keep the recession going well into 1981.	2. How would an inventory liquidation keep the recession going?

In the first three months of 1980, Canadian factories were operating at 84.6 per cent of capacity, compared with 88.4 per cent 15 months earlier.

The reason for this fall in output is, of course, a build-up of inventories as demand falls away.

3. Explain why a build-up of inventories causes a fall in output.

Economists expect to see some further liquidation of inventories this quarter, but at a slower rate. And that change from rapid to slow—from negative to less negative—shows up as a plus for the economy.

4. Why would this change be a plus for the economy?

But as consumer spending slackened, stocks began to pile up. So an inventory correction is underway, and it will reduce this quarter's real gross domestic product by more than 10 per cent at an annual rate.

5. What is an "inventory correction"? Explain how it would "reduce this quarter's real gross domestic product."

Until last week, the weight of evidence seemed to be on the side of a rapid cooling-off this winter of the strong growth posted in the last three months of 1987. A sharp downturn in consumer spending in October and an enormous build-up in inventories in December caused many economists to predict the five-year-long, consumer-led recovery in the U.S. was running out of gas. On top of that was January's surge in unemployment insurance applications.

6. a) Is the build-up in inventories consistent with the downturn in consumer spending?
   b) How would you portray the build-up in inventories, and the downturn in consumer spending, on a 45° line diagram or on an aggregate supply-aggregate demand diagram?

The Commerce Department has just reported that manufacturers' stock of raw materials, goods in process, and finished items fell by 1 per cent in May, the latest month for which figures are available. It was the biggest drop in 17 years.

But it isn't just the sheer size of the May decline that excites the experts. Rather, it is the fact that the figure extends a trend that has been going on for most of the year beyond their previous expectations. The cumulative impact—on top of earlier large work-downs of retailers' holdings—has now reached the point where it has become a more potent force for recovery.

7. The first paragraph talks about an inventory change, while the second paragraph talks about the inventory level. For each, explain how this information affects your prediction of the future level of economic activity.

C. Determinants of Investment

Business capital spending is providing a lot of the fuel for the economy this year and will be a major source of strength in 1989 as well. This is typical of the mature stage of the business cycle. Demand has reached levels where capacity constraints are appearing—the manufacturing capacity utilization rate came in at 80.3 per cent in the first quarter, with seven of the 21 major sectors reporting rates above 90 per cent. And companies have enough cash to finance new investment, with after-tax corporate profits up 35 per cent last year. As a result, quarterly growth rates for business capital spending are likely to be in the 1–2 per cent range through 1989.

1. What two determinants of investment demand are put forward in this clipping?

Slower consumer spending is expected to be offset by strong business investment, spurred in part by capacity constraints. Capacity utilization in manufacturing hit a five-year high of 81.7 per cent in the fourth quarter.

2. What are capacity constraints and how can they spur investment demand?

"When manufacturers lost confidence in the mid-1970s, they stopped investing, and now they have to make up for lost time."

That loss of confidence followed a loss of cost competitiveness by Canadian industry. But things have turned right around, with lower wage increases, a lower Canadian dollar, lower government growth and higher rates of capacity utilization, he said in a recent interview.

3. The last sentence of this clipping cites four reasons why investment has been stimulated. For each of these phenomena, give a brief explanation of why investment has been encouraged.

Both these factors may tend to prolong the coming recession. Companies with close computer control of inventories will be able to empty their warehouses much more thoroughly before kicking the production machine into gear again. And the cost of maintaining an inventory at today's lending rates will give them a powerful motive to do their best.

4. a) Do you agree that computer control of inventories will lead to longer recessions? Why or why not?
   b) What relationship does the article suggest exists between inventories and interest rates? To which category of demand, C, I, or G, would this be related?

Weak markets will lead to a rapid drop in the rate of business capital spending during the next year, which in turn will increase unemployment.	**5.** a) What major determinant of investment demand is focused on here? b) Explain the logic behind the conclusion that unemployment will increase.
The modest firming of investment intentions over the past half-year took place in the face of a looming recession, which now seems to have arrived in Canada as it clearly has in the United States. The momentum survived despite record high interest rates earlier this year, and it is being maintained at a time when the federal government's demands on the capital market are unusually high. Fears had been expressed that financing of the federal deficit— which will require the raising of $11-billion to $12-billion in cash this year—might squeeze private borrowers out of the market. Fortunately, corporations have come through a period of sharply rising profits and are in a good position to finance their capital outlays.	**6.** a) By what mechanism does federal deficit financing "squeeze private borrowers out of the market"? b) What three major determinants of investment spending, one positive and two negative, does the article make reference to?

D. Investment Impact and Timing

If the trade deal is scuttled after the federal election, business investment will likely drop—putting a sharp brake on the economy as a whole.	**1.** Explain how this brake will operate.
With consumer spending showing signs of a slowdown, hopes for continued economic expansion are now pinned on business investment in new plants and machinery.	**2.** Explain how business investment could continue economic expansion.
The capacity utilization rate is not only a measure of how busy industries are, but also gives some indication of future employment and investment moves.	**3.** Explain how the capacity utilization rate can give "some indication of future employment."

Recession fears may also be compounded by the traditional wisdom that base metals surge at the end of an economic cycle. The rationale is that only at the tail-end of a growth period do capital spending and expansion kick in—increasing the demand for metal products.

4. Why does investment demand (capital spending) kick in at the end of a growth period?

But if that suggests the economy is on the upswing again, history indicates that nothing, perhaps, could be farther from the truth. Though welcome to those who service them, major surges in capital spending occur almost only when the national economy is approaching or is in recession.

5. Why would "major surges in capital spending occur almost only when the national economy is approaching or is in recession"?

II-4 Fiscal Policy

The two major macroeconomic policies used by governments to affect the operation of their economy are monetary and fiscal policy. Monetary policy, discussed in the next section, concerns government manipulation of the money supply in the economy through the Bank of Canada's control over chartered banks. Fiscal policy concerns government manipulation either of tax rates or of its own level of spending. Pure fiscal policy, as distinct from monetary policy, refers to changes in government spending or taxing that do not involve concomitant changes in the supply of money. This type of government policy is usually the first to be introduced to students, via a shifting of an aggregate demand curve, with much being made of the subsequent multiplier effects. The favorite example is an increase in government spending; a tax decrease is another popular example. Both lead to an increase in economic activity through a chain of reactions called the multiplier process. An increase in government spending, for example, increases aggregate demand for goods and services, causing a rise in income. This in turn increases consumption, further increasing demand. This causes another increase in income, perpetuating the process. When this multiplier process has worked itself out, the level of income will have increased by more than the original increase in government spending.

Economists first recognized the potential of fiscal policy after the publication of Keynes's *General Theory* in 1936, but it was not until the early 1960s that they were successful in selling this potential to government leaders. Now, all budget speeches include reference to the impacts of changes in government spending and taxing, and to their multiplier effects. These basically Keynesian ideas, embodied in all introductory economics texts, can also be found in most newspaper articles dealing with macroeconomic policy, as the short clippings below illustrate.

Monetary policy (discussed in the next section) is also used to stimulate the economy, and has similar multiplier effects. But monetary and fiscal policies differ in several fundamental ways. For example, one policy may be easier to enact than the other, or may affect the economy more quickly or more reliably, or they may have differing impacts on variables such as interest rates.

One major difference between them is the extent to which they are discriminatory and the extent to which that discrimination can be controlled. Monetary policy affects the housing industry and small businesses more strongly than other sectors, a discrimination that is very difficult to alter. A potential advantage of fiscal policy, on the other hand, is that its discrimination can be easily controlled, allowing the policy to be used to affect particular sectors of the economy (specific geographical or industrial sectors, for example).

II-4A Using Fiscal Policy

Defence spending in U.S. brings down unemployment

by Ruben Bellan

WINNIPEG—Fifty years ago the *London Economists* published an article titled "Guns to the Rescue." Its theme was that Britain's heavy spending on armaments, promoted by the growing threat of Nazi Germany, was creating jobs for workers long unemployed. The production of guns was rescuing the economy from the clutch of prolonged depression.

Several years before the government had declared itself to be absolutely unable to finance a housing program that would have provided both jobs and badly needed housing. Now it was spending far larger amounts for weapons of war and thereby generating jobs for many thousands of workers in arms factories.

Arms Contracts

Guns are coming to the rescue in the U.S. today. Ronald Reagan's immense armament program has been exercising an ever-increasing effect on the national economy. More weapons contracts are being let, more munitions plants are being built, completed weapons factories are adding to their staffs, and sub-contractors are gearing up to supply components and materials.

Literally millions of American workers are now employed in armament production: they are generating employment for millions more as they spend their earnings on housing, clothing, entertainment, eating out and so on. U.S. arms production on a gigantic scale may endanger the world but it certainly is creating jobs; it is unquestionably the major reason for the substantial decline in the U.S. unemployment rate.

Other forms of spending by the U.S. government could equally be creating jobs. Millions of Americans need better housing. All kinds of public works projects should be carried out.

The fact that the U.S. is experiencing economic recovery primarily because of spending on armaments reflects the mindset of its policy makers. They are alarmed at the size of the federal budget deficit and firmly determined to hold down government spending in order to keep down the deficit. They are willing to accept an increase in government spending only for a purpose which is transcendingly important.

Canadian policy makers share American economic ideology. They too are convinced of the need to hold down the federal deficit and of the need therefore to keep federal spending down. They do not, however, share Reagan's enthusiasm for acquiring weaponry. The consequence is that Canada's unemployment rate is remaining at double-digit levels while the U.S. rate is going down significantly.

Committed socialists contend that the free enterprise system depends on war and preparation for war to achieve full employment. They are wrong: it doesn't. As Canada has demonstrated in the past and as Japan and a few western countries are demonstrating today, it's perfectly possible for a free enterprise country to keep its labor force fully employed in the production of goods for peaceful use.

Public Works

What is needed is a national government which, whenever unemployment looms, undertakes needed public works and services and provides new stimuli to the country's private business firms.

If the Canadian government spent enough on new public projects and sufficiently boosted the private sector it could generate jobs on a very large scale, as the arms program is doing in the U.S. Canadian business leaders are strongly opposed, however, insisting that the government cannot afford large expenditures for such purposes. No small irony is involved. The country's most ardent supporters of the free enterprise system vehemently insist on policies which would prove the Socialists to be right.

Ruben C. Bellan is professor of economics at St. John's College, University of Manitoba.

1. Which passage reflects the multiplier process associated with this U.S. fiscal policy?
2. What role does the budget deficit play in influencing fiscal policy? Does the author view this as a positive or a negative influence? Explain.
3. Which is the boldest statement of this author's belief in
   a) the feasibility (and desirability) of having the government undertake discretionary fiscal policy, and
   b) the strength of fiscal policy?
4. Explain the logic behind the final sentence.

Short Clippings

Note: Suggested answers for the odd-numbered questions below appear in the Appendix.

A. Government Spending

There are specific reasons, of course, why a downturn may have been aborted in the past few months. With the Russian invasion of Afghanistan and the long confinement of American diplomats as hostages in Iran, both the U.S. and Canada began to strengthen the armed forces. Defence contracts work in the old-fashioned Keynesian way to stimulate the economy.

1. Explain how this "old-fashioned Keynesian way" operates to stimulate the economy.

Under present circumstances there *is* such a thing as a "free lunch." In effect, all of the ingredients of that lunch are already there—the people who want to work, the factories and equipment standing idle, and the raw materials not being used. It is just a question of injecting a little spending power to prevent the free lunch from going to waste.

2. What kind of government policy is this a plea for?

But in a loose sense, we are all Keynesians now—all of us, at any rate, who reject the notion that a sick economy heals itself by "natural" recuperative powers, without government action.

3. Explain why rejecting this notion makes us Keynesians.

In the last six months of 1979, less than 40 cents of every dollar of production in Canada was passing through government hands.

That's the first time the government share of the economy dipped below the 40 per cent level since 1974.

In 1979, according to the GNP figures just issued by Statistics Canada, the government share of the economy dropped back to 40.3 per cent, and averaged 39.9 for the last six months of the year.

The federal share was down to 15.6 per cent, from 16.6 in 1978. The provincial share was down to 12.3, from 12.7 per cent. And the municipal share was down to 8.6, from 8.9 per cent.

4. a) Do these figures imply that the G (in the well-known relation Y = C + I + G) in 1979 was 40.3 per cent of GNP, 15.6 per cent of GNP, or is it not possible to tell from the information given? Explain.

b) This clipping uses the terminology "share of the economy." In what sense is this correct, and in what sense is it misleading?

For Canada, the demand pressures generated by the U.S. tax cuts and the spillover effects of increased U.S. defense spending will push the Canadian economy further into an excess demand situation.

5. a) What kind of U.S. macroeconomic policy is referred to here?
   b) Why will this U.S. policy affect Canada?

As for expanded spending, a lid should be kept on it. There are enough programs so that the federal deficits will swell as existing programs are maintained while tax revenues fall as a result of the weak economy. The expanded deficit will automatically become stimulative.

6. What technical terminology do economists use to describe this phenomenon?

With savings up and private sector demand for money down, it is almost a classic case of room for government stimulus without damage.

7. What is meant here by government stimulus "without damage"?

B. The Multiplier

What is needed for the economy now is increased government spending and a cut in taxes so that people have more expendable income. With more income on hand, people will tend to spend a percentage of this new income which in turn will increase others' incomes.

1. What technical terminology do economists use to describe a) this policy action, and b) the reaction described?

Instead, he suggested a radical idea: governments could jump-start a return to prosperity by running deficits. Rising consumer demand would then lead to higher profits, more investment, and a renewed cycle of growth. This brilliant and simple insight has dominated economic thinking for decades.

2. a) What name is often given to a policy of running a deficit?
   b) What name is given to the process generating the "renewed cycle of growth"?

For the people around here it means 42 direct jobs and, depending on the employment multiplier you prefer, another 120 or 160 indirect jobs. In the bush. On the booming grounds. On the highways. In the drugstore down the street.

3. Suppose the "it" in the first sentence refers to an increase in spending of $1 million.
   a) What is meant here by the term "employment multiplier"?
   b) Calculate the magnitude of this multiplier.

"Smith's $14,040 in wages brought the three levels of government some $4,800 in direct and indirect taxes. When Smith spent his $14,040 on goods and services, he provided income for the suppliers of those goods and services and for the suppliers of these suppliers; the income of all these suppliers brought the three levels of government some $3,400 more in direct and indirect tax revenue, for a total government revenue of $8,200."

4. From this information, what do you calculate the multiplier to be?

- **Reduced multipliers:** the evidence has been available for some time that real multipliers have fallen and may now be below one.

  This means that government deficits are no longer stimulative: an increased deficit of $1 billion will yield less than a $1-billion increase in real output and will accelerate inflation.

5. a) What is the difference between a multiplier and a "real" multiplier?
   b) Which would you expect to be larger? Why?
   c) Show how this difference can be portrayed on an aggregate supply-aggregate demand diagram.
   d) In what special case of this diagram will the two multipliers be the same?

C. Tax Policy

Taxes should be cut substantially if the government really intends (as it said in last week's throne speech) to make "a serious assault upon high priority goals . . . to stimulate industrial expansion, put more Canadians back to work."

There have been times when the federal government kept pumping fiscal stimulus into the economy when it was already growing vigorously and did not need the boost. The result was. . . .

1. a) How does economic theory explain how cutting taxes will stimulate industrial expansion?
   b) Complete the last sentence.

This leads some economists to see tax-cutting, strange remedy as it may seem, as the first step to a more balanced budget.

2. Explain the logic behind this statement. Wouldn't cutting taxes reduce tax revenues and thus make the budget deficit worse?

If Mr. MacEachen de-indexes income tax deductions, he will be introducing huge, annual, automatic tax increases so long as inflation lasts.

3. Explain why de-indexing would be equivalent to tax increases.

A few years back the finance ministry calculated the relative benefits of a tax cut to stimulate consumer spending and concluded it should use its money in other ways.

The example cited was of a $1 million cut in the sales tax as compared to spending $1 million on public works. Each measure would cost the provincial treasury the same amount, but the construction project would create 33 worker-years of employment and return $160,000 in revenue to the province, while the sales tax cut would create only 25 worker-years of employment and return just $118,000 in increased revenue.

4. Explain why this report is either consistent or inconsistent with economists' theoretical views of the balanced budget multiplier.

The OECD data also indicate, for example, that the stimulative effects of tax reduction are considerably smaller than those triggered by government expenditure increases—a fact of life long recognized by economists.

5. Explain why this is so.

•**Indexing**: indexing of personal income taxes was intended to take inflationary gains out of tax revenues. It was also intended to be a discipline on governments.

6. Explain how indexing was supposed to act as "a discipline on governments."

D. Budget Deficits

If it were a "stimulative deficit" which was deliberately undertaken to prime the pump of the economy, then the debt might liquidate itself.

1. Explain the mechanism by which a deficit could "prime the pump" of the economy. How might this cause the debt (caused by the deficit) to liquidate itself?

What cannot be done, various reformers in the U.S. notwithstanding, is to impose on any government the obligation to balance its budget annually. Consider the consequences. If it did work, it would introduce a major destabilizing element.

2. Explain how forcing the government to balance its budget would be destabilizing.

"But governments have got themselves into this position over a long period of time so they can't just go out and slash the deficit," he said. "It would be nice if they could but that would put too much of a brake on the economy and that's not what we need right now. (Deficit reduction) has to be a long-term thing."

3. Explain how cutting the deficit could put a brake on the economy.

Adam Smith said it 200 years ago: "What is prudence in the conduct of every private family can scarce be folly in that of a great kingdom."

With the advent of a "Keynesian revolution" since World War II, principles of fiscal responsibility were abandoned—in fact, reversed. The message of Keynesianism might be summarized as: "What is folly in the conduct of a private family may be prudence in the conduct of the affairs of a great nation."

4. a) Exactly what is the "folly" referred to in this clipping?
   b) Explain the rationale behind the "message of Keynesianism."

The Keynesian analysis, which has dominated economic thinking for the past half century, made government deficits respectable.

5. How did Keynesian analysis do this?

"I think a lot of people, particularly in the business community and among economists, have realized that a lot of the deficit that we're knowing at the present time is of a purely cyclical nature.

"It's resulting from what they call the automatic stabilizers; that is. . . ."

6. Complete this explanation.

A federal deficit is supposed to stimulate the economy and a bigger deficit is supposed to stimulate more than a smaller one, but this test is far from perfect.

7. Why is this test far from perfect?

E. Miscellanea

And as soon as the budget is produced, we're likely to hear a lot about the danger that government borrowing will "crowd out" private borrowing.

1. Explain how this crowding out is brought about.

In assessing the IMF data, our conclusion is that Canada could not withstand as "painlessly" as the U.S. the significant fiscal drag required to bring about a big reduction in the federal deficit over the next couple of years.	**2.** Explain the meaning of fiscal drag and what impact it would have on the economy.
A second reason for questioning the employment-generating abilities of the deficit involves how consumers respond to an increase in government borrowing. Rational consumers realize that a higher deficit now means higher taxes in future. If people. . . .	**3.** Complete this explanation of how the behavior of rational consumers could lead to crowding out.
Another criticism of defense spending is that it promotes inflation by adding to total demand in the economy without increasing the supply of new goods available to meet it, because a soldier spends his pay on things he has not helped produce and the munitions worker does not go out and buy a tank.	**4.** Comment on this argument.

II-5 Monetary Policy

The money supply is controlled by the Bank of Canada, mainly through open-market operations—the buying or selling of government bonds. Purchases of existing government bonds by the Bank of Canada should be sharply distinguished from its purchases of new bonds, however. Existing bonds are held by the public; their purchase by the Bank means that the populace is left holding more money but fewer bonds—the same level of wealth is held, but in a different form. Purchase by the Bank of new government bonds, by contrast, gives the government money which it then spends, increasing the money holdings of the populace with no corresponding decrease in their holdings of bonds—their total wealth holdings are larger. This latter action on the part of the central bank is often called "printing money"; because of the concomitant increase in wealth, this means of increasing the money supply is much more stimulative than is the former means. The inflationary impact of a government deficit often results from the fact that the deficit is financed by selling bonds to the Bank of Canada rather than to the public.

By affecting the economy's money supply, the Bank of Canada also affects interest rates. The most prominent manifestation of this is the Bank rate, regarded by markets as a bellwether of future Bank of Canada policy. This rate is set at a quarter-point above the average yield on three-month bills sold at the Thursday treasury-bill auction. By buying more or fewer T-bills, the Bank of Canada can influence the outcome of this auction. Prior to 1980 the Bank rate was just "announced" by the Bank of Canada; the post-1980 policy (reviving a 1956–62 policy) is referred to as a "floating" Bank rate.

In spite of the fact that Keynes himself stressed the importance and potential of monetary policy, fiscal policy was adopted by the "Keynesian" economists as "the" policy tool, and monetary policy was relegated to a position of minor importance. It wasn't until the 1960s that a revival of interest in monetary policy occurred. This was the result of two things. First, a number of empirical studies appeared, indicating that monetary influences on the economy were considerable and were, in the view of the more controversial studies, more powerful and reliable than fiscal influences. And second, the predictions of the "monetarists" (the defenders of the view that monetary actions are more important than fiscal actions) in the 1960s proved more accurate than those of the "Keynesians."

This revival of interest in monetary policy spurred examination of the way in which monetary policy affected the economy. In the Keynesian view, an increase in the supply of money lowers the interest rate (a rise in the supply of money, relative to its demand, causes the "price" of money, the interest rate, to fall) and the lower interest rate stimulates spending (consumers will buy more durables, with a lower cost of credit; business will find more investment projects profitable since the cost of borrowing is less; and municipal and provincial governments will find it financially feasible to build more ice rinks or schools). In the monetarist view, this mechanism is supplemented by general considerations of portfolio equilibrium. An increase in the money supply means that on the whole people in the economy have a wealth portfolio with relatively more money (than other assets) in it than before. To rectify this imbalance, they will draw down their money balances by buying other assets such as consumer durables (directly increasing aggregate demand) or equities (increasing their price and thus lowering the implicit price of capital goods, increasing investment demand).

In the 1970s interest in monetary policy accelerated as inflation accelerated. Ever since recorded time man has noticed that increases in the money supply and increases in prices have gone hand-in-hand. This was spectacularly so in hyperinflations such as in Germany after World War I, in the American Confederate States near the end of the U.S. Civil War, and in many South American countries today. It was also the case in less spectacular instances such as in the 17th century when the Spanish discovered gold in the New World and in the second half of the 18th century when fractional-reserve banking was invented. True to form, the inflation of the 1970s was accompanied by large increases in the money supply; in Canada the money supply in some years reached annual increases exceeding 20 per cent. Although increases in government spending or cost-push forces can create an inflation as easily as can an increase in the money supply, only an increasing money supply can *sustain* inflation. Any inflation involves an ever-increasing demand for money to undertake transactions at higher and higher prices; without an increase in the money supply this higher demand for money will force the "price" of money, the interest rate, higher and higher, eventually causing it to reach heights sufficient to curtail aggregate demand and grind to a halt those forces pushing prices up. It is this consequence of a fixed money supply which explains why governments allow it to grow and thus sustain the inflation: Governments do not like to have high interest rates and high unemployment.

This phenomenon is formalized by the textbook result that in the long run an economy's level of inflation is equal to the difference between its money supply growth and its real income growth. If real income is growing at 3 per cent, for example, then the demand for money will also grow at 3 per cent. A 7 per cent growth in money supply will imply an extra 4 per cent of money that people will not wish to hold as money. In the simplest variant of this theory, people "get rid of" this extra money by spending it, pushing up prices. If prices rise by 4 per cent, the demand for money rises by 4 per cent, eliminating the excess supply of money and thereby stopping the price rise; an inflation of 4 per cent results. This is the thinking that lies behind the monetarist policy rule: Increase the money supply at a fixed rate equal to the real rate of growth of the economy.

The monetarists' persistent advocacy of this policy rule has caused considerable controversy, usually summarized in textbooks under the heading "policy rules versus discretion." For example, it is not clear what definition of the money supply should be employed in operationalizing the rule, it suggests that central bankers are not to be trusted and should be replaced by robots, and it prevents monetary authorities from reacting to obvious money demand shocks such as postal strikes, stock market crashes, or the failure of a major financial institution. All this provides fertile ground for newspaper commentary.

II-5A T-Bill Auction

Average yield of federal bills climbs to 11.17%

Dow Jones Service
OTTAWA

The average yield at this week's auction of $2.8-billion of 91-day Government of Canada treasury bills was 11.17 per cent, up from 10.95 per cent last week.

As a result, the Bank of Canada's bank rate moves to 11.42 per cent from 11.20 per cent.

Accepted bids for the 92-day bills ranged from a high of $97.300 for an 11.13 per cent yield to a low of $97.288 for an 11.18 per cent yield. The average bid price was $97.290.

(Reprinted by permission of Dow Jones News Service, © Dow Jones & Company, Inc. 1989. All Rights Reserved Worldwide.)

Note: Suggested answers to the questions below appear in the Appendix.

1. The first sentence states that the average yield is up from last week. What does this imply about what has happened to the average bid price?
2. How do they get the 11.42 per cent figure cited in the second paragraph?
3. Explain how one would calculate the result that a bid of $97.300 corresponds to a yield of 11.13 per cent.
4. Suppose the Bank of Canada had not wanted the bank rate to rise as much as it did. What action would the Bank have taken to ensure a smaller rise?

Struggling Monetarism

Why 'monetarism' is struggling

By Andrew Coyne

IT IS increasingly widely believed, among laymen and academics alike, that "monetarism is dead"—not just that it is out of vogue, but that as a basis for economic policy it is no longer practicable or even desirable.

It is not so much monetarist objectives that have fallen into disrepute, as it is the particular means by which they are to be achieved: that is, of targeting particular measures of the money supply as a means of controlling economic activity. This certainly appears to be the thinking at the Bank of Canada. Whether the reports of monetarism's death prove greatly exaggerated will depend on whether this technical problem can be overcome.

The central core of monetarist thought, as mentioned, remains firmly lodged in the mainstream of economics, and can be summarized as follows:

☐ The growth of total output in dollar terms ("nominal GNP") is determined by the interaction of the supply and demand for money. Since the demand for money tends, over time, to follow a predictable path, governments may control the growth of nominal GNP by an appropriate regulation of the money supply.

☐ Increases in the money supply cannot, however, alter the economy's real productive potential. While monetary stimulus may for a time push real growth above that potential, eventually it affects only prices.

☐ The money supply should therefore increase only enough to accommodate growth in potential output, but since that is difficult to gauge, and since money increases take effect with unpredictable lags, monetary policy will do least harm by keeping money growth on some fixed rate.

☐ More than that, recent theory stresses, it should be *seen* to be fixed: if economic players know what the central bank is up to, their actions and expectations of inflation will adjust more readily; inflation will come down more rapidly, and with less drop in output. Central banks should therefore set explicit, well-advertised monetary targets. The closer the authority hews to the monetary rule it sets for itself, the less will output deviate from trend.

This is where the difficulty starts. What Western countries attempting to apply monetarist prescriptions have discovered is that the accelerating pace of innovation in financial markets has given the public many new ways to hold money. This has blurred the distinction between different measures of the money supply, as individuals and businesses shift into and out of financial assets, and has made individual measures increasingly meaningless as indicators of overall monetary conditions.

The example of the Bank of Canada is instructive. Most people will recall that Bank of Canada Governor Gerald Bouey introduced monetary targeting to the Canadian scene in 1975, picking M1—currency plus noninterest bearing chequing accounts—as the money supply that mattered.

In the early years of setting ranges for M1 growth, the money supply came in more or less on target, and its relationship with total spending in the economy remained more or less predictable.

In 1981–82, however, things began to go very wrong. The relationship between the money supply and output and prices became far less certain. While the growth rate of M1 dropped off alarmingly, nominal GNP did not trace a similar path (see chart).

What caused this decoupling was a substantial drop in the demand for M1. This was largely due to the introduction in 1982 of daily interest chequing accounts, which attracted a growing portion of funds away from ordinary chequing accounts, and increasingly from savings deposits.

The upshot was that a given amount of M1 supported a higher level of total spending, or in other words, its velocity increased—not, as had been the case during the years of accelerating inflation in the late 1970s, along a predictable trendline, but at a markedly greater pace.

At which point, the Bank of Canada threw up its hands. Rather than adjusting the target, or possibly using a different money measure, the Bank, in November, 1982, gave up targeting altogether. Despite assurances at the time, it has shown no signs of returning to the practice; Bouey's annual report, issued last month, makes no mention of targets.

The result has been the destruction of any semblance of a stable expectational climate. It is today very difficult to tell what the Bank's current monetary policy is. Rules are out; ad hocery is in. Says University of Western Ontario Professor

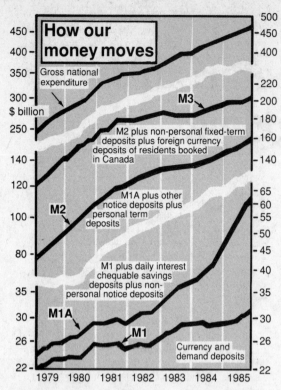

Bank of Canada Governor Gerald Bouey abandoned his policy of targeting M1 in 1982. Now he looks at all money measures.

David Laidler: "They appear to be looking at everything from week to week and not telling anybody what they're doing." This uncertainty was surely a factor in the run on the C$ in 1985–86, since there was no guarantee that Canada's deficit burden would not be relieved by monetization.

What's a central bank to do? A recent issue of the Bank of Canada Review, surveying developments in monetary aggregates over the past few years, speaks glowingly of the performance of M2 as an indicator, which suggests the Bank is at least giving it increased prominence in its thinking.

There is good reason for this: since M2 encompasses currency, ordinary chequing accounts, and interest-bearing chequing and savings accounts, shifts between each component will not affect its overall level. A broader measure, M2+, which includes deposits at the increasingly important nonbank financial institutions, performs even better as an indicator.

Whether M2, M2+, or any of the 43 other measures the Bank's research department has devised are guiding Bank policy, however, without a more forthright public declaration of fealty to some monetary rule, the Bank's commitment to price stability lacks credibility. Further lasting progress against inflation, therefore, will come at a greater cost than necessary in forgone output and employment.

---

1. According to this article, the main reason for the "death" of monetarism is that one of its central beliefs (stated in the fourth paragraph) is not true. Which belief is that?

2. In the sixth paragraph it is claimed that "the money supply should therefore increase only enough to accomodate growth in potential output." What will happen if the money supply grows by, say, three percentage points more than this? Explain.

3. a) What is the definition of M1 velocity, as introduced in the thirteenth paragraph?
   b) What has caused the increase in M1 velocity?
   c) Why would the accelerating inflation of the late 1970s have increased M1 velocity?

4. In the fourteenth paragraph it is stated that the Bank of Canada "gave up targeting." Exactly what is this "targeting," what is it supposed to accomplish, and how is it supposed to work?

5. What is meant, in the fifteenth paragraph, by the statement that "there was no guarantee that Canada's deficit burden would not be relieved by monetization"?

6. Near the end of the article the possibility of using M2 as an indicator is suggested. What is it about M2 that immunizes it from the criticism of M1 introduced in the eleventh paragraph?

Short Clippings

Note: Suggested answers for the odd-numbered questions below appear in the Appendix.

A. **Changing the Money Supply**

Money deposited for a term is not left in bank vaults but is loaned out by the banks (subject to minimum cash reserve requirements). This means that a dollar on deposit can flow back into the banking system one or more times and that dollar can expand money supply.	1. Explain the meaning of the last sentence.
The deficit will mainly be financed through the sale of securities to the general public and not to the banking system.	2. What are the implications of selling the securities to the public and not to the banking system?
"It's true we're still not cutting into credit expansion, which is the source of all our problems," he said.	3. How can credit expansion be "the source of all our problems"?
Though Bouey doesn't say so, this may ultimately compel him to resort increasingly to managing the money supply by managing chartered banks' excess cash reserves—the stuff from which the banks create loans. That's a supply-minded approach to the problem, as opposed to a price-minded approach, which followers of Nobel-Prize winning economist Milton Friedman have long advocated.	4. a) What are "excess cash reserves"? b) How would they be "managed" by the Bank of Canada? c) Explain the difference between a "supply-minded approach" and a "price-minded approach" in this context.
Indeed, the central bank has surely considered this possibility and has already attempted to assuage fears that there might be a terrific surge in money supply as proceeds from the bond issues are redeposited with chartered banks.	5. a) Explain how this terrific surge would come about. b) Why would the proceeds be redeposited with the chartered banks anyway?

In the process, the money multiplies, since the banks are allowed to lend more money than they actually have, within limits set by the board. The board tries to anticipate how much the money will multiply as this process unfolds. If its calculations are right, just enough money will be created to accommodate the growth it desires for the economy. If the calculations are wrong, a monetarist board would make them right by pumping more money into the economy or pumping some out.

6. a) What "money" (in the first line) is being multiplied here?
   b) What terminology is used to refer to the magnitude of this multiplication?
   c) What are the "limits set by the board"? (The board is the Federal Reserve Board.)
   d) Why is the adjective "monetarist" employed in the last sentence?

B. The Bank Rate

One analyst's swift response was that Bouey and his political masters, by allowing the bank rate to float with market rates, have escaped much of the political flak attending unpopular interest rate increases, particularly as the Canadian economy's growth slows. Indeed, Bouey has said he dislikes the controversy surrounding each painful announcement of bank rate increases.

1. a) Explain how floating the bank rate will allow Bouey to escape political flak.
   b) Why is this escape only cosmetic?

It is fixed each Thursday afternoon at a quarter-point above the average yield on three-month federal Government treasury bills offered at the regular weekly auction in the money market.

There were indications the increase in the average bill yield would have been greater if the Bank of Canada had not participated heavily in the auction.

2. a) To what does the first word of this clipping refer?
   b) Why would the Bank of Canada participate in this auction, and how would it cause the increase in the average bill yield to be less?

Why? Probably because of the lacklustre change in policy which now ties the central bank rate to the yields of 91-day treasury notes.

Make no mistake. The Bank of Canada has not abdicated its role as a setter of interest rates, it has only changed the manner in which it accomplishes the change.

3. This clipping states that although the bank rate is tied to treasury bill yields, the Bank of Canada still sets interest rates. Explain the new "manner in which it accomplishes the change."

Yesterday, the bank appeared to make no attempt to influence the results of the weekly auction of federal government treasury bills in the money market. It sets its trend-setting lending rate at a quarter-point above the average yield on three-month bills sold at the auction.

Earlier in the week the bank's traders intervened aggressively in the money market to push the yield on last week's bills sharply higher.

4. a) How would the Bank influence the results of the weekly auction?
   b) Exactly what kind of intervention is being referred to in the last sentence?

Secondly, it should be noted that the bank itself can—and does—control the T-bill rate. Last week, bank traders held the 90-day T-bill average at 13.68 per cent only by feverish purchases of bills. One trader, who doubts the bank received any bids for bills under 13.75 per cent, suggests the bank may have swallowed the entire half a billion dollars of bills.

5. a) What are the implications for the interest rate of the Bank's "swallowing the entire half billion dollars of bills"?
   b) What are the implications for the money supply?

C. Measuring Money

The annual growth rate in what had once been the officially watched aggregate, M1 (currency in circulation plus demand deposits) was 4 per cent from one year ago, 4 per cent over a six-month period and 1.9 per cent over the past three months.

For M1A (M1 plus daily interest chequable and nonpersonal notice deposits), the annual growth rate was 12.1 per cent over the past 12 months, 13.2 per cent over six months and 14.6 per cent over three.

1. Why might M1A be growing so fast relative to M1?

The Bank of Canada's statement said the latest weekly figures show that the rate of growth in the narrowly defined money supply, M-1, moved below its current target range of five per cent to nine per cent annual growth during May.

It identified three causes: the delayed impact of rising interest rates earlier this year; more moderate growth of spending in the economy; and the increasing use of daily interest savings accounts by bank customers.

2. Explain in your own words how each of the cited "causes" contributes to a lowering of the growth of M1.

Another potential thorn in the flesh of central bank money-measures is the daily interest savings account.	3. Why would daily interest savings accounts be a "thorn in the flesh of central bank money-measurers"?
Because they are broader in coverage than M1, M2 and M2+ are less prone to the shifts resulting from financial innovation that caused difficulties in the interpretation of M1.	4. Explain what these difficulties were.
The impact on this monetary aggregate of extensive financial innovation—the changes in the kinds of deposits and services offered by banks that occurred in the early 1980s—led the Bank of Canada to drop M1 as an intermediate target in 1982. With the changes in the way the public was holding payments balances, the M1 aggregate no longer had the same reliable link to. . . .	5. Complete this clipping.
In a recent interview, Bank Governor Gerald Bouey told *The Post* the chances of monetary targeting returning are "not very good, at least in the short run because of the shifts going on in the banking system as a result of bank policies and technology."	6. a) What is "monetary targeting"? b) Why would "bank policies and technology" prevent the Bank of Canada from returning to monetary targeting?
On the latter point of technology, Bouey said two factors are emerging: First, corporations can almost wind up the end of the day with a zero balance so that there are very little transactions balances. For individuals, the trend to chequable savings accounts is firmly established with a mixture between savings and transactions balances (remaining in their accounts). "It's won't be clear as to the breakdown between the two functions for which the account is held," he said.	7. Of what relevance is the point raised in the last sentence?
These technological advances make it possible to keep more money in interest-bearing accounts. And the level of interest rates makes it not only attractive but almost a necessity. The loss of income on money kept in cash and demand deposits is now so large that it is worthwhile to scheme to keep even small amounts earning interest.	8. What does this clipping suggest concerning the influence of the interest rate on the demand for money? Is your answer consistent with textbook theory?

174 Part II *Macroeconomics*

Bouey does acknowledge that M1 can be a slippery commodity. Probably since 1976, the Bank—in its weekly tally of the money supply—has significantly underestimated the transactions money flying around the system.

Around that time, the chartered banks' bigger corporate customers cottoned onto the advantage of so managing their moneys as to achieve, at a consolidated, central current account, practically flat balances.

For the chartered banks promoting this switch of idle money to easily accessible interest-bearing deposits, of course, there were similar economies: reserve requirements on such deposits are much lower.

9. a) What is "transactions money" (as opposed to other kinds of money)?
   b) Why would corporations want to achieve the flat balances noted in the second paragraph?
   c) What is the relevance of the comment in the last paragraph that "reserve requirements on such deposits are much lower"?

D. A Monetary Rule

Of course, a large body of economists now believes that if the 1970s taught us anything it's that macroeconomic fine-tuning is more trouble than it's worth. In effect, the economy's steering gears are so loose that unless you hit a very sharp curve, fiddling with the wheel is just as likely to do harm as good.

1. What kind of monetary policy is this an argument for?

Far better for central bankers to get out of the fine-tuning business. Instead, they should try to keep . . .

2. Finish this clipping.

Bouey states that there "are no aggregate measures or indicators of the rate of monetary expansion in Canada that are sufficiently reliable at present to be used as targets for policy, or that are uniquely helpful in the task of explaining the impact of monetary policy."

Accordingly, judgment about financial and economic conditions—rather than following a monetary rule—will continue to be the guiding force behind monetary policy. Thus, it would appear that a return to monetary targeting—the system which the Bank nominally used between 1975 and 1982—isn't in the cards.

3. a) What kind of "monetary rule" is being referred to here?
   b) How would such a rule work and why is it being abandoned?
   c) What kind of "judgment about financial and economic conditions" would be relevant here?

Running an economy is a balancing act. It's frustrating to have to keep a whole lot of things in balance and it would be nice if it could be reduced to a simple formula such as that of the monetarists.

There are difficulties in having the central bank present growth rate targets in advance. Central banks find it much easier to explain events after the fact if they are allowed to operate from a position of secrecy, with little communication with the public. Politically unpopular increases in interest rates might also have to be tolerated at embarrassing times under a policy of published target growth rates.

The wiser course would be to avoid switching monetary policy and keep the money supply growth stable. This will automatically bring interest rates down during the recession when demand for funds is weak.

Whatever the solution, the failings of monetarism, even if they are overcome, are a matter of concern to non-monetarists and monetarists alike. Monetarism's recognition of the usefulness of monitoring the money supply was an important contribution to economics, they say. And they accept the principal that, eventually, an economy that goes on producing more money than non-inflationary growth requires will turn inflationary.

"In the long run, monetarism has to be correct," said Richard Darman, the deputy secretary of the Treasury. Excessive money growth has to lead to inflation.

"The problem with monetarism," he said, "is that its advocates have seen it as infallible over short periods of time and wish it to be rigid in its application over all periods of time."

4. a) What is the simple formula referred to?
   b) What is one thing this simple formula does not prevent from moving about considerably? Explain.

5. Why might increases in interest rates have to be tolerated under a policy of target growth rates?

6. Explain the rationale behind the last sentence.

7. a) To what textbook result does the last sentence of the first paragraph refer?
   b) Explain the mechanism by which "excessive money growth has to lead to inflation," as stated in the second paragraph.
   c) Explain in your own words, with an example, the problem with monetarism stated in the last paragraph.

E. Monetary Policy

Monetary and fiscal policy are the two big "levers" by which an economy can be controlled.
　Monetary policy, in the hands of Gerald Bouey of the Bank of Canada, is mainly an other name for the manipulation of interest rates. Fiscal policy takes its name from the "fisc" or public treasury in ancient Rome. It's about tax rates, government spending and deficits.

1. What alternative definition of monetary policy is possible? Which is preferable? Why?

The question always seems the same: Is the Fed tightening or easing? But analysts approach it through different methods. Some focus on interest rates while others focus on the growth of reserves. And, as with much in economics, these two approaches can give different answers.

2. Explain how it would be possible to get two different answers by using these two approaches.

"John Crow has got it right," says Merrill Lynch's Manford. "His job is to take the punch bowl away just as the party gets roaring. And if you have 6 per cent real growth a quarter like we did last year, it's quite a party you've got going."

3. What is the "punch bowl" in this context?

Several bond issues will raise $400 million, with the Bank of Canada picking up at least $200 million.
　The Bank of Canada can also be counted on to take down more of the bonds than the planned $200 million it has announced if they seem to be selling badly.

4. a) What are the implications of the Bank of Canada's picking up $200 million of these bond issues?
　b) What does the second sentence imply about the Bank of Canada's monetary policy?

Most major governments appear willing to stick to a non-accommodating, tighter monetary policy, "without rushing to the Keynesian spigot." But there is not unanimity on this view. With elections coming up in Germany, France and the United States, the more cynical expect some kind of pump-priming policies in all three countries.

5. a) What is a "non-accommodating" monetary policy?
　b) What is the meaning of "rushing to the Keynesian spigot"?
　c) What is a "pump-priming policy"? Explain the meaning of the name.

"The thrust of the Bank of Canada's policy has been to reduce demand for loans on the basis of price—not by limiting the amount of money that is available."

"It isn't the function of the chartered banks to be self-appointed rationers of credit. We are in a very competitive business; if my bank won't lend to you, another one will."

Those two comments summarize the responses of senior bankers who were asked by *The Province* whether their present lending policies—particularly on consumer loans—are consistent with the central bank's objective of lowering the inflation rate by cutting back on credit demand.

6. a) Are reducing demand on the basis of price and limiting the amount of money available separate phenomena?
   b) Do the banks ration credit? Explain how or why not.

This is the bottom limit of the range of 5 per cent to 9 per cent growth that the Bank believes is healthy for the economy.

However, with the sharp fall in interest rates, Bank officials expect a pick-up in economic activity that will bring about an upsurge in monetary growth.

7. This clipping suggests that the lower interest rates will cause the money supply to increase. Isn't the causal relationship the reverse; i.e., doesn't a change in the money supply cause a change in the interest rate? Explain the meaning of this statement.

The eagerly awaited weekly money stock figures measure more than just the supply of money. The weekly number measures money demand as much as it does supply.

The evidence now suggests the alarming growth rate of the money stock mainly reflects an upsurge in money demand, rather than an overly expansive money supply policy.

8. How can this be explained?

178 Part II *Macroeconomics*

F. Regional Policy

Mr. Crow has said lower interest rates should not be used to encourage growth in the regions because the economy in Central Canada is close to overheating and lower rates would add to . . .	**1.** Complete this clipping.
Expecting a national monetary policy to fulfill regional needs, to Beigie's mind, is like hammering a square peg into a round hole. "What Crow is not saying, but what he is implying," says Beigie, "is that you are only going to get regional policies through . . .	**2.** Complete this clipping.
There could only be one monetary policy for Canada, said Crow, not separate ones for each province or region. And the policy has to be set for the majority of Canadians, who just happen to live in central Canada. A less diplomatic man might have added that the onus for developing and implementing regional policy lies not with the Bank of Canada, which controls monetary policy, but with . . .	**3.** Complete this clipping.
We see only the latest in a long series in the conflict between the high interest rates necessary to cool Ontario's booming economy, and the low interest rates the rest of us still would like to have to just warm up a bit. The Bank of Canada, having no regional levers, does what it has to do. The rest of us just suffer the results.	**4.** a) Explain why the Bank of Canada has "no regional levers." b) What policy is needed here?
For example, the strength in house-building activity was concentrated in Central Canada while activity dropped off in British Columbia and Alberta. The economic disparity has also created concerns that the high level of interest rates will hurt the weak regions of the country while failing to cool those regions of the economy that are overheating.	**5.** Suggest a means whereby policy makers can address this dilemma.

The proper lines of action for controlling inflation are not too difficult to envisage. First, we must stop relying on monetary policy to do the job. The monetary tool has been easiest and most convenient for all governments to use because it requires no legislation; it's a matter of men speaking in well-modulated voices around a polished table with charts on the wall. It has been given a good trial. But it hasn't worked, and its economic impact has been discriminatory.

6. How has monetary policy been discriminatory?

Monetary policy is a peculiar tool in that it does not seem to have an intermediate switch: it is either completely ineffective; or it is too effective. In the past few months, it has been too effective, almost destroying the housing industry in the United States and nearly paralyzing the Canadian housing industry.

7. How would monetary policy destroy the housing industry?

II-6 Inflation

Inflation is defined as a persistent rise in the general price level. Although conceptually everyone "knows" what inflation is, from having to deal with it in their everyday lives, few realize the problems involved in its measurement. This is the subject of the first article below.

Monetarists believe that "inflation is always and everywhere a monetary phenomenon"—that an inflation can be created and sustained only by excessive money creation. (This reflects our discussion of the monetarist rule in Section II-5.) Consequently, they prescribe a drastic reduction in money supply growth as the cure for inflation. In 1975 the Bank of Canada espoused this monetarist view and adopted a policy known as "gradualism"—decreasing the rate of growth of the money supply gradually over several years. Unfortunately, this policy did not meet with as much success as the Bank of Canada had hoped, prompting Governor Bouey himself to note in retrospect that the Bank's policy may have been "too gradual."

In contrast to the hard-line monetarists, many other economists believe that, in addition to excess money supply growth, there are other major forces playing important roles in the inflation process and that these forces should not be ignored when structuring policies to fight inflation. Cost-push forces are an important example of this. As production costs rise firms increase prices. These price increases cause everyone in the economy to try to raise their wages or the prices of the goods or services they sell to ensure that they don't suffer a fall in their standard of living. This starts the process anew, and if it is "accommodated" by money supply increases, it will continue the inflation. Wage increases due to union power play an important role in this process, but increases in the price of basic commodities, such as energy, have become the most important ingredient in recent years.

Notice that these cost-push forces only maintain the inflation if they are accommodated by money supply increases, allowing the monetarists to claim that the real "cause" of the inflation is the money supply increases, not the cost-push forces. This gets one into a semantic quarrel concerning the real "cause" of inflation. In our view, the ultimate "cause" of inflation is what is causing the monetary authorities to increase the money supply at an excessive rate. A discussion of this is beyond the scope of this book; a summary can be found in Chapter 10 of P. Kennedy, *Macroeconomics*, 3rd edition (Boston: Allyn and Bacon, 1984).

This section cannot be viewed as a comprehensive look at inflation; it is a bridge from Section II-5 on monetary policy to Section II-7 (interest rates), II-8 (international macroeconomics), and I-9 (stagflation). These sections address the relationship between inflation and interest rates, exchange rates, and unemployment, respectively.

II-6A Measuring Inflation

Here's how they peg the consumer price index

The Economy
Don McGillivray

OTTAWA—What is the consumer price index?

In the third week of every month Statistics Canada announces the consumer price index for the previous month.

The number everyone wants to see is not the CPI itself but how much it has increased in the past 12 months. This is reported as the "inflation rate" and usually is taken as a fever-chart of the economy.

Where does Statistics Canada get the CPI? What's left out? And does it really measure the cost of living for the ordinary Canadian?

Here are some answers:

What is the consumer price index? The main index is a number showing how prices are changing, on average, for families and people living alone in 64 Canadian cities with populations of 30,000 or more. In other words, it measures consumer costs for city-dwellers, not for farmers or people in small towns.

How does Statistics Canada get the CPI? It's based on a "basket" of goods and services bought by the ordinary city family. Goods are such things as tomatoes, gasoline, shoes and furniture. Services are things such as haircuts, dry cleaning and an evening at the movies.

StatsCan sends its shoppers out on a pricing trip each month, checking the costs of some 400 items. In all, they collect about 100,000 prices. These are fed into the StatsCan computer that produces the CPI.

What is an index? It's a special kind of percentage that is easier to handle than the raw numbers. StatsCan could just add up all the prices—so much for soap, so much for cornflakes, so much for milk, and so on. This would result in an amount expressed in dollars and cents. But this would be hard to compare.

So StatsCan says, in effect, "Let's say that all these prices equalled 100 in 1981. That's the base period. Now let's see how much more than 100 they would come to now."

The result is an index that reaches 118.5 in August, 1983. In other words, the average Canadian city dweller was paying $118.50 last month for what he or she could have bought for $100 in 1981. That's an increase of 18.5 per cent.

And 5.5 per cent of this increase was in the past 12 months.

How does StatsCan know that is what the average family actually bought? It doesn't. But it bases the basket on a family-spending survey, last taken in 1978, and tries to "weight" the various items so they're counted as the right proportion of total spending.

Housing, for example, is now counted at 35.4 per cent of the average family's budget. Food takes about 21 per cent. Transportation, including such things as bus fares and the cost of running a car, is the next biggest part at 16 per cent. Clothing takes nearly 10 per cent. Recreation, reading and education are grouped together at 8.6 per cent. The other main items are tobacco, alcohol, health and personal care.

Does the CPI adjust when consumers switch to less costly foods? It doesn't, and that's one main reason StatsCan insists that it measures consumer prices, not the "cost of living."

Suppose the cost of beef goes skyhigh. Many people will switch to chicken or pork. But to make its CPI valid from month to month, the agency has to stick to the same basket of goods. So the CPI assumes people buy as many steaks when prices are high as when they're low.

This means most people can "beat the CPI" by switching to cheaper goods.

What's left out of the CPI? It's supposed to measure consumer prices, so things that aren't consumed are omitted. An example is the land on which your house stands. Rising prices for houses are in. Rising land prices aren't.

The CPI includes sales taxes that raise the price of goods. But it doesn't include increases in income taxes—although most families count them as a cost—because they affect how much money you have to spend, not the cost of the things you buy.

Is the CPI a fair measure of inflation? Not entirely, because it only measures rising consumer prices. If wages or interest rates or income taxes or the prices companies have to pay for raw materials are rising rapidly, that's inflation, too. But it will show up in the CPI only when it affects what consumers pay.

Is the CPI so flawed that it's useless? Not at all. It's a rough measure of inflation as seen through the eyes of the average citizen. As StatsCan itself said in a defence of the index last year: "Don't shoot the CPI—whether bringing good news or bad, it's the best messenger we've got."

Note: Suggested answers to the questions below appear in the Appendix.

1. Explain in your own words the difference between a price index and a cost-of-living index.
2. Do changes in the price index understate or overstate changes in the cost of living? Explain.
3. Suppose the "basket" of goods and services in 1978 cost $680 and in 1986 cost $952. If the corresponding price index had a value of 100 in 1978, what value would it have in 1986? What would be the average annual rate of inflation during this eight-year period?
4. Under the title "What is an index?" the author suggests that we "could just add up all the prices," whereas in fact the prices are not all given equal weight when they are added up. What weights are used and how would they be obtained?

II-6B Inflation Fight

Inflation fight not yet won

AS A FEDERAL election draws nearer, the prospect of an economy awash in money becomes likely. It is therefore reassuring that John Crow, Governor of the Bank of Canada, remains vigilant about the dangers of inflation.

In a comprehensive speech last week at the University of Alberta (a digest appears on this page), Crow made clear his goal is nothing less than price stability. That means, as he indicated to the Commons finance committee later in the week, *zero* inflation—not the 4 per cent-plus at which we have been stalled these past four years.

But talk is one thing; policy persuades. If zero inflation is to be achieved without substantial cost to the real economy, it is more critical than ever that public expectations be locked on the same wavelength as monetary policy. Otherwise, businesses and workers will go on basing decisions on the 4–5 per cent inflation rates they are used to, and price themselves beyond demand.

Clear, well-advertised money-supply rules do more than a hundred pronouncements to convince the public of the authorities' anti-inflationary resolve. It is here that Crow's speech appears to mark an important departure: the governor came as close as possible to declaring a return to monetary targets, without actually doing so.

The question is why he did not go all the way, and set preannounced, percentage growth rates for the broad money aggregates. Crow made plain the Bank has been considering various money-supply aggregates as possible targets ever since it abandoned the narrow-measure M1 in 1982. But the experience of M1 still unnerves: innovations in the banking system, especially interest-bearing chequing accounts, allowed the public to shift money holdings into and out of different instruments in such a way as to make the relationship between M1 and total spending in the economy—its "velocity"—unstable.

The broader-measure M2, however, since it encompasses most significant forms of money individuals may hold, resolves the problem. The shifts in asset holdings take place *within* the broader measure. That doesn't make it impervious to distortion, but M2 velocity has shown substantial predictability for several years.

A monetary rule need not mean a single, bald number: if Crow fears velocity shifts, he could set out rules of like clarity for adjusting the target in the face of a trend change in velocity. The governor likens the conduct of monetary policy to "driving in a rainstorm with defective windshield wipers." Give us some well-lit signposts.

---

1. What is the relationship between "the prospect of an economy awash in money" and inflation, suggested in the first paragraph?
2. What does "price themselves beyond demand" (third paragraph) mean? Why would it be a problem?
3. What are the "monetary targets" referred to in the last sentence of the fourth paragraph, and of what relevance are they to inflation?
4. a) How is velocity measured?
   b) Explain how a change in velocity could affect inflation.
5. The second-last paragraph suggests that the "predictability" of velocity is important. Explain why.
6. Explain the relevance of the last sentence.

Short Clippings

Note: Suggested answers to the odd-numbered questions below appear in the Appendix.

A. Measuring Inflation

He asked Industry Minister Jack Horner to explain why the relative weight of food was being reduced in the new index in spite of recent rapid increases in food prices.

1. How would you explain this action?

A spokesman for the employer's council who asked to remain unidentified said the CPI is "far from perfect. It's an imprecise measure, but it's the only one available."

The index is a frequent factor in negotiations, but it does not reflect individual consumer patterns, it assumes everyone buys new products and it ignores the fact people substitute cheaper goods for more expensive ones.

2. Explain how each of the three items mentioned in the second paragraph affect the interpretation of the CPI.

The bank estimates that the annual inflation rate in January would have been only 3.5 per cent rather than 4.1 per cent had it not been for indirect tax increases imposed in 1987.

And that does not include any income tax increases.

3. a) Explain the logic behind the first sentence.
   b) How would income tax increases affect the measured inflation rate, as suggested in the last paragraph?

The rapid development of computers "makes it difficult to determine how much of the change in measured prices of computers is due to pure price change or due to the change in quality of the product," Statscan said. "It is clear that the cost of obtaining a given amount of computing capability has been falling."

4. a) What is the relevance of the change in quality?
   b) What does the last sentence imply about how computer prices should affect the CPI?

B. Causes of Inflation

But, increasingly, there does seem to be agreement on one point, namely, that the advancing share of governments in the economy has been strongly inflationary. This is mainly because individuals and companies have not been prepared to pay (through taxes) for extra public services or income redistribution and have, instead, demanded bigger and bigger income increases to offset the tax and keep real incomes growing.

1. The argument presented here is not the usual argument claiming government deficits are inflationary. What is the usual argument? Are these arguments demand-pull in nature or cost-push?

Q: Higher rates of interest must boost the cost of all sorts of things for businessmen, whether they need loans to finance inventories or because the nature of their business entails time-lags between delivering the goods and getting paid. How can higher interest rates be anti-inflationary?

2. How would you answer this question?

"There is little doubt that the key reason for this slide toward ever more inflation was an effort by public policy in most countries to achieve and maintain more output and employment from their economies than was consistent with price stability."
—Gerald Bouey

3. a) How would public policy be used to attempt "to achieve and maintain more output and employment"?
   b) Explain why this could lead to inflation, as Bouey's statement implies.

There's nothing wrong with any level of interest rates produced by the natural forces in the economy. But artificially lowered rates are dangerous because they are produced by artificial increases in the money supply. And the extra money sluicing around in the system at a time when production is close to capacity is more likely to go into higher prices than anywhere else.

4. How would you define what is meant here by "artificial" increases in the money supply?

"I wouldn't say that a lower dollar is inflationary. A lower dollar means you're paying more for your imports. In other words, it adds to your cost of living and it can become inflationary. . . ."

5. Complete former Prime Minister Trudeau's statement to explain how it can become inflationary.

C. The Role of Money

The first group, known as monetarists, have the virtue of being simple—brutally simple, some say. The theory is that prices will expand to use up all the available spending money in the country. Thus, for zero inflation what you have to do is keep money supply growth to the rate of increase in output or real GNP.	1. Explain the rationale of this monetarist view in your own words, so as to make it more understandable.
If, as monetarists believe, too rapid growth of money supply is a prime cause of inflation, then increases of about 20 per cent a year seem alarming.	2. What is meant here by "too rapid"? What rate would be about right?
Inflation results when the quantity of money and credit is increased, on a sustained basis, at a rate that is higher than the economy's capacity to produce real goods and services.	3. Explain how this causes inflation.
Winning the battle but losing the war? That seems to be the situation of the Bank of Canada in its fifth year of trying to apply the monetarist prescription to defeat inflation by squeezing the money supply.	4. Explain how inflation can be defeated by the monetarist prescription of squeezing the money supply.
How can this be? How can the economy show just as much inflation with money growing only half as fast? After all, the growth of M1 has been cut to about seven per cent, which is close to the five per cent that Bouey said four years ago would go with "the restoration of a completely stable price level, that is, a zero rate of inflation."	5. a) How would one calculate the money supply growth rate consistent with "the restoration of a completely stable price level," claimed by Bouey to be five per cent? b) Answer the question posed in the first two sentences.
The rapid expansion of M2 and M3 occurred as inflation rates were at historically high levels. Instead of saving, people spent and turned money over quickly in the banking system.	6. Why would money tend to turn over more quickly (high velocity) during an inflation?

D. Unemployment Implications

The Governor warned about a possible "collision" in the future between rising wages and prices and his hold-the-line monetary policy.

1. Explain the nature of this collision and the results it will produce.

This suggests that wages play a game of catch-up, then fall behind for a while. But on the average, they advance about three per cent a year, in real terms. We are still not far off this pace.

The only trouble is that the economy is no longer productive enough to yield three per cent average wage gains in real terms. If the cycle now goes into the catch-up phase, the demands for real gains to make up for the real losses of the past three years will collide with an economy going into at least a temporary stall.

2. Explain in terms of economic theory how this "temporary stall" will come about and what will happen to cause it to be only temporary.

Just as in the '30s when government intervened to save capitalism from itself, so again must it intervene massively now to save government capitalism from itself. Since government now protects individuals and corporations from the consequences of excessive wage and price increases (that is, under old-style capitalism, of personal failure), government must now prevent those excessive increases by permanent wage and price controls.

3. Exactly how does the "government now protect individuals and corporations from the consequences of excessive wage and price increases"? What are these consequences?

"The Bank of Canada was presumably trying to maintain very high interest rates and, I assume in the face of political reality and pressure from the Canadian economy, released the interest rate a few weeks ago," he said.

"The monetarists will allow you to go ahead and ruin people and countries but when eventually in good and common sense you say 'enough is enough' the monetarists say 'well, you spoiled the experiment.'"

4. Explain exactly what the "experiment" is and how and why it was spoiled.

E. Miscellanea

It's ironic that the monetarist policy of the Bank of Canada and the Trudeau government was intended to prevent runaway inflation from shaking public confidence in the economic system.

But part of this policy has been to destabilize interest rates. And this effect was deliberate, although perhaps not intended to go as far as it has.

1.
a) What is the monetarist policy referred to?
b) How does it prevent runaway inflation?
c) How does it destabilize interest rates?

"The message from the bank to the private sector is: Do not give in to inflationary psychology, do not assume that you can easily accommodate cost pressures, do not expect monetary accommodation of accelerating hydrocarbon price increases."

2.
a) What does "assume that you can easily accommodate cost pressures" mean?
b) What does "monetary accommodation of accelerating hydrocarbon price increases" mean?

In addition, the more popular versions of monetarism do not take into account the fact that, as incomes rise and interest rates fall, most people want to hold larger cash balances than when income growth was almost nil and interest rates were in double digits. As a result, the monetarist fears that the strong U.S. money supply growth of the past year would ignite inflation have, so far, proven unfounded.

3. Explain in your own words the logic of this argument.

That governments increase the money supply at a too-rapid rate because it enables them to finance an expanding government sector without the requirement of increasing taxes or borrowing from the economy's real savings.

4. What does borrowing from the economy's real savings mean?

His solution to Canada's economic troubles? Allow inflation to run at a rate of about 8 or 9 per cent, reduce interest rates "and much if not all the problems will disappear."

Skalbania, who was caught in the collapse of real estate prices three years ago, reached an agreement in January, 1983, with his creditors.

5.
a) Why would this policy be attractive to a real estate speculator like Nelson Skalbania?
b) What's wrong with this policy?

II-7 Interest Rates

Everyone worries about the interest rate: home buyers and owners because it determines the size of their mortgage payments, businesses because it affects the cost of expansion, and governments because voters don't like high interest rates. This leads to frequent newspaper commentary on interest rates and speculation on their future levels.

In the financial sections of newspapers, interest rates are most frequently mentioned in discussions of the bond and money market. The main reason for this is the inverse relationship between interest rates and bond prices—when interest rates rise, bond prices fall, and when interest rates fall, bond prices rise. Suppose you pay $100 for a bond which pays 10 per cent. Each year you "clip" a coupon worth $10 off this bond, and when it matures you are paid back your $100. Now suppose the interest rate rises to 11 per cent (i.e., bonds currently being issued are paying 11 per cent). It is to your advantage to sell your 10 per cent bonds and use the proceeds to buy 11 per cent bonds. But no one will want to pay $100 for your 10 per cent bond because they can get 11 per cent elsewhere. To sell your 10 per cent bond you will have to drop its price, in this case to a price low enough that the $10 coupon (plus the ultimate rise in the value of the bond to $100 on its maturity date) creates a yield of 11 per cent. This explains why, when interest rates rise, bond prices fall; the process would operate in reverse if interest rates were to fall. We have already seen this phenomenon in action in the section on monetary policy; by undertaking open market operations (buying or selling bonds), the Bank of Canada affects the price of bonds and thus changes the interest rate.

It should now be evident why participants in the bond and money market are so interested in the interest rate: If interest rates rise, holders of bonds will suffer a capital loss, and if interest rates fall they will enjoy a capital gain. Accurate forecasts of future interest rates can therefore be very profitable.

Forecasting interest rates requires careful analysis of the determinants of interest rates. By far the most important determinant is the expected rate of inflation. Suppose the interest rate is 5 per cent and that suddenly everyone expects prices to rise by 3 per cent (instead of zero per cent) during the next year. Those loaning money at 5 per cent will expect to receive, at the end of the year, dollars that will buy 3 per cent less than they would at the beginning of the year, so their expected net return is only 2 per cent. To obtain a return of 5 per cent they will want to charge 8 per cent. Those willing to borrow earlier at 5 per cent should now be willing to pay 8 per cent because they expect to save 3 per cent by buying their car (for example) now rather than next year. In this example, 8 per cent is called the "nominal" or "money" rate of interest and is the rate that appears on the market and in the newspapers; 5 per cent is the "real" rate of interest, obtained by subtracting the expected rate of inflation from the nominal rate. Because the expected rate of inflation cannot be measured, the real rate of interest is an unmeasurable quantity. For a long time, the real interest rate hovered in the 3 per cent range, but during the 1980s it appears to have risen to the 5 or 6 per cent range, reflecting a change in individuals' preference for present versus future consumption, and government borrowing to finance large budget deficits.

All this suggests that a prediction of the future rate of inflation (based, for example, on measures of the rate of increase of the money supply) is the primary ingredient of a forecast of future interest rates. Other determinants of the interest rate are also taken into account. The level of income affects the demand for money, so that, for example, if the economy moves into a recession the demand for money will fall (relative to its supply), lowering the interest rate. Movements into recession are usually accompanied by a fall in inflation, so expectations of inflation are likely to fall, supporting any decrease in interest rates. Furthermore, we might expect the monetary authorities to try to stimulate the economy in a recessionary period, doing so by increasing the money supply and lowering interest rates. Thus speculation concerning central bank money-supply behavior plays a prominent role in the forecasting of interest rates.

Interest rates are also affected by forces arising from Canada's economic interaction with other countries; this is discussed in the next section.

II-7A Challenging Conventional Wisdom

The Capital Market

by Hugh Anderson

Money Supply

The conventional wisdom says that cutting back the rate at which money is created will push up interest rates. But economist David Herskowitz of A.E. Ames and Co. took a look at the actual experience during 1980 and found that the truth seems to be the other way round.

He measured money-supply growth each month in the United States and Canada, on a three-month moving average basis, and compared the numbers with representative short and long term interest rates and with the year-over-year change in the consumer price index—for the first 11 months of the year.

"In both the United States and Canada interest rates were lowest during the year when money supply growth was slowest in June and July," Mr. Herskowitz reported in the firm's weekly Money and Bond Markets letter.

"As money supply growth speeded up in Canada and in the United States in the second half of the year, interest rates increased."

Mr. Herskowitz recalled that the Secretary of the U.S. Treasury, G. William Miller, said recently that the Federal Reserve Board plans to slow money supply growth in 1981. Mr. Miller suggested this would "drive interest rates through the roof."

Mr. Miller "may be right that money supply growth will slow this year, but this suggests interest rates will decline rather than rise," Mr. Herskowitz commented.

Note: Suggested answers to the questions below appear in the Appendix.

1. Explain the logic behind the "conventional wisdom" cited in the first sentence of this article.

2. Use economic theory to explain the evidence found by Herskowitz.
3. With respect to the last two paragraphs, do you agree with Miller or Herskowitz? Why? How would you reconcile these conflicting viewpoints?

II-7B Investment Planning

Deflation may be key to investment planning

BONDS

BY JOHN GRUNDY

The outlook for inflation is the main consideration for bond market participants who are devising near-term investment strategies. It is generally conceded that the trend of inflation has a major influence on the market and overshadows other determinants of the present worth of fixed-income investments.

Widespread concern that inflation is in an uptrend is backed by reasonably convincing evidence and seems echoed by the policies of major central banks. Recently, most official price measures have confirmed that monetary policies designed to counter the potential for much higher inflation are justified.

In most industrialized countries and particularly the United States, current economic conditions are typical of those usually preceding an acceleration of inflation. The supply of labor is tight, many industries are operating at close to maximum capacity, consumer demand while slowing remains high, and import costs are rising.

Under such circumstances, it is prudent for potential bond market investors to adopt a cautious approach, particularly to longer-term issues. But because the market tends to discount probable near-term developments, it is equally important to consider if current prices allow for increased future inflation—or whether the concern has been overdone.

A small but growing base of market analysts believes that the latter is the case; that inflation pressure, at least for this economic cycle, has passed its peak and the potential for disinflation or deflation should now dominate market participants' strategies.

If the economy slows or moves into a recession phase, inflation and inflation expectations will fall. And while general business conditions are more buoyant than many analysts had predicted, over-all growth has steadily subsided in recent months.

There are now fewer bottlenecks being reported in the manufacturing pipeline, which leads to faster deliveries and tends to dampen the pressure for higher prices. And fewer companies are currently planning to raise prices, as producer costs have levelled off in recent months.

Greater competition should also hold the rein on inflation, and general economic expansion, as it gets older, will become more fragile. This in turn will lessen upward price pressures and may encourage price reductions as manufacturers scramble to maintain sales.

Neither is the recent growth in business loans from the banking system conducive to higher prices. Loan extension by the banks has slowed as they attempt to repair their portfolios from the damage of past abuses. Partly offsetting revenue from traditional loans has been the growth of fee income from such sources as payment guarantees and establishing lines of credit to corporate customers.

Such activities do not lead to an increase in the money supply. Unless money grows at a faster rate than is needed to maintain the economy's momentum, the fuel needed to feed inflation is not present.

1. Explain what is meant in the second sentence by "the trend of inflation has a major influence on the market."
2. Explain how the trend of inflation determines the present worth of fixed-income investments, as noted in the second sentence.
3. Explain what is meant by "monetary policies designed to counter the potential for much higher inflation," described in the second paragraph.
4. Explain how the economic conditions listed in the second sentence of the third paragraph would be reflected on an aggregate supply-aggregate demand diagram.
5. Why is the phrase "particularly to longer-term issues" added to the first sentence in the fourth paragraph?
6. Exactly what is meant by "current prices allow for increased future inflation," as stated in the fourth paragraph?
7. With respect to the comments made in the fifth paragraph, what strategy should bond market participants adopt if they believe that inflation will fall in the future?
8. What have business loans got to do with higher prices, as claimed in the second-last paragraph?
9. What textbook result is being appealed to in the last paragraph?

Short Clippings

Note: Suggested answers to the odd-numbered questions below appear in the Appendix.

A. **Bond Prices and Yields**

Lenders would be wise to move their short-term holdings into longer-term assets to take advantage of the capital gains generated by declining interest rates.	1. a) What are the capital gains mentioned in the clipping, and how do they come about? b) Why aren't these capital gains associated with short-term as well as long-term assets?
Yields on treasury bills started to rise early in the week as dealers began to unload the bills because of higher than anticipated carrying costs.	2. How would dealers unloading bills cause their yields to rise?
Basically, investors profit in two ways from putting their money into bonds: through the coupon rate, or interest rate, that is attached to each bond and provides a steady income; and through potential . . .	3. Complete this clipping.
For example, a 20-year Government of Canada bond paying 10.25 per cent was selling for $1,007.10 Nov. 25. The bond was priced above the par value of $1,000 to yield 10.19 per cent—a rate competitively close to those offered in the U.S.	4. The quoted yield of 10.19 per cent is slightly lower than $102.50 expressed as a percentage of $1007.10. Why?
Capacity utilization at 82.4 per cent, was unchanged in January for U.S. mines, factories and utilities. Economists are worried that demand on industry may soon outstrip capacity, thereby encouraging producers to raise prices. Inflation, the undisputed Achilles heel of bonds, would result.	5. Explain why inflation is the "undisputed Achilles heel of bonds."
It was a year of mixed economic signals, according to Marc Meagher, chief economist at Merrill Lynch Canada Inc. "Swings in the price of the bellwether Canadian bond reflected a high degree of uncertainty about inflation and changing perceptions in the market," he reports. "The release of figures that showed the economy to be stronger than expected tended to push prices _____ and yields _____ and vice versa."	6. Fill in the two blanks here and explain your reasoning.

196 Part II *Macroeconomics*

A smaller-than-expected decrease in the U.S. money supply dealt the North American capital market a hard blow, as bond prices sagged across a broad front.

7. Explain why bond prices sagged.

B. Bond Markets

One view in the market has been that because the U.S. economy seemed weaker than it should be, the U.S. central bank, the Federal Reserve Board, would cut its discount rate from 7 1/2 per cent. This, of course, would be positive for bonds.

On the other hand, there have been those thinking that rising money-supply growth would rule out such a discount rate reduction.

1. a) Why would a cut in the discount rate "be positive for bonds"?
   b) Why might rising money-supply growth rule out a discount rate reduction?

. . . the shock of Thursday's flash second-quarter news that the U.S. economy has grown *three whole percentage points*.

You and I would say that's good news. But the bond markets' terrified interpretation last Thursday was that it might encourage the private sector to borrow, nudging up interest rates. Add that discomforting prospect to the other horrifying disclosure—that, at last reading, U.S. money supply had climbed by a mammoth US$4.8 billion—and you'll know why people were heading for the U.S. and Canadian bond market exits.

2. Explain in your own words why people were heading for "bond market exits."

In the financial markets, red ink is flowing on Bay Street as some investment dealers continue to lose money on the treasury bills, notes and bonds they buy from the federal Government.

The Bank of Canada, unwilling to discourage a major market for its debt, has helped the dealers out over the past week or so, taking unwanted three-month treasury bills off their hands in exchange for one-month bills, thus reducing their risk in the marketplace.

The biggest buyers of the bills, the chartered banks, have been reducing their holdings steadily since last September in order to meet rising loan demand.

3. a) Why are investment dealers losing money?
   b) How does the Bank of Canada's action "reduce their risk in the marketplace"?
   c) How is the information conveyed in the final paragraph consistent with the first two paragraphs?

Falling interest rates in both Canada and the U.S. have triggered a rally in the bond market that has many investors rejoicing.

4. How would falling interest rates "trigger a rally in the bond market"?

Analysts say the three-month rally in bonds has been fuelled by Washington's promise to balance its budget by 1991 and OPEC's decision to abandon support for world oil prices in the short run.

5. Explain the logic behind these two explanations for the bond rally.

And many borrowers fear that if North American inflation can't be whipped, Canada's bond markets may evolve into replicas of Europe's capital markets—surrendering their status as providers of long-term, fixed-cost funds to government and industry.

Of course, some economists—notably Wood Gundy Ltd.'s John Grand and Burns Fry Ltd.'s Leigh Skene—see long-term bonds returning to favor with a vengeance as inflation finally succumbs to slowing money-supply growth. Others, though, envisage inflation and interest rates peaking higher and higher each business cycle, as sound money management succumbs to electoral politics, so that investors progressively shorten their commitments.

6. a) Explain why the bond markets may have to surrender "their status as providers of long-term fixed-cost funds."
b) What does "sound money management succumbs to electoral politics" mean, and why would it occur?
c) Why would "investors progressively shorten their commitments"?

NEWS OF U.S. economic weakness last week cleared the way for higher bond prices.

Traders went on a buying spree after Friday's announcement by the U.S. Labor Department of a lower-than-expected increase in nonfarm payrolls.

The Canadian and New York markets moved quickly to capitalize on this good bad news; prices on both markets shot up more than a point ($10 on every $1,000 of face value) in minutes.

7. a) Why is this referred to as "good bad news"?
b) Explain why news of economic weakness would lead to higher bond prices.

And the slow-growth idea is also bullish for the bond markets because lower growth means low and falling interest rates and this in turn pushes up the value of fixed-income securities.

8. Why does lower growth mean low and falling interest rates?

"The recent surge in the money supply must be contained, and in very short order, if the hope for a viable long-term bond market is to be sustained," said John Grundy of Toronto-based Commercial Union Assurance Co. of Canada.

9. a) What has the money-supply surge got to do with the viability of the long-term bond market?
   b) Why is the modifier "long-term" used?

C. Real versus Nominal

It should be noted, though, that Dungan does not expect real interest rates to be as low as Ip does, predicting a real rate of around 4 per cent in 1989 (a prime of about 8 per cent and a CPI rise of 4 per cent).

1. Explain how the real interest rate prediction follows from the prime and CPI rise predictions.

Mr. Rogalski said: "The yield of every fixed-income instrument is determined by two things, expectations of the inflation rate and expectations of real return. This deal takes the uncertainty out of the inflation component."

The Franklin Savings bonds will carry a 3 per cent premium over inflation.

2. a) What real rate of return will an investor get if he or she buys this inflation-indexed bond?
   b) Will one be better off buying this bond (rather than a regular bond) if the current market expectation of inflation is too high or too low? Explain why.

It became all too evident that market interest rates embody expectations about future inflation. And when strong expectations of high inflation develop, interest rates that otherwise look high are not really so in regard to their impact on economic behavior.

3. Does this clipping suggest that the real or the nominal interest rate is the primary determinant of economic behavior? Explain.

As a result, the argument runs, interest rates, which particularly affect certain sectors (capital spending, houses and cars) are raised to levels which in previous times were viewed as impossible. But clearly, they aren't offside with what the central bank views as appropriate.

4. Explain the argument that would explain how interest rates could rise to "levels which in previous times were viewed as impossible," and would still be "what the central bank views as appropriate."

Moreover, contrary to Bouey's pronouncements, higher interest rates have not caused everyone to borrow less. Most people are borrowing as much as ever—or more—to buy goods now, because they think that prices will be even higher in the future.

5. What explanation would you offer for the fact that higher interest rates have not caused a drop in borrowing?

D. Inflation and Interest Rates

In the light of the devastation brought about in the 1970s by severe inflation, world financial markets have tended to be extremely sensitive to any signs of an increase in price pressure. Any fears of a pickup in inflation have led quickly to a _____ in long-term bond rates.	**1.** Fill in the blank here and explain your reasoning.
Bouey said last month of Canadian interest rates that "the extent to which we can get them down and keep them down depends mainly on getting the rate of inflation down." Of course, one of the problems with this prescription is that high interest rates are themselves inflationary as they feed into costs and prices. So the cure is part of the cause of the disease.	**2.** a) Explain the rationale behind the statement in the first paragraph. b) How would you rebut the argument given in the second paragraph?
Corporate treasurers should not be frightened by the recent rise in interest rates on bonds. Rates of even 13 per cent will look like bargains if inflation heats up over the next 18 months. Investors should continue to shun the market for long-term securities.	**3.** a) Explain why rates of even 13 per cent will look like bargains. b) Why should investors shun long-term securities if inflation is predicted to rise?
The realistic alternative to high interest rates is not lower interest rates (except temporarily) but even higher interest rates and a monetary inflation that would make our heads spin.	**4.** a) Explain why a policy of lower interest rates would lead to "higher interest rates and a monetary inflation that would make our heads spin." b) Why has the qualification "except temporarily" been included?
"There is no doubt whatsoever," Bouey's deputy, R.W. Lawson, said in a Calgary speech last November, "that the net impact of rising interest rates on a market economy is anti-inflationary." The trouble with this reasoning is that people no longer believe an increase in interest rates is a sign that inflation is about to be beaten. Quite the opposite. People now take a jump in interest rates as a sign that inflation is about to get worse.	**5.** a) Why is the adjective "net" employed by Lawson when talking of the impact of interest rates on inflation? b) Is there any logic behind the view expressed in the last sentence, or is it purely a psychological phenomenon? Explain.

"A decline in the rate of inflation is the one sure route to lower interest rates." So said Bouey in his annual report for 1981, published in March, 1982.

Inflation is only about one-third of what it was two years ago, but interest rates are higher. How come? We've paid the price, a fearsome price in slow growth and high unemployment, for "the one sure route to lower interest rates." Why haven't we had the promised results?

He said the whole exercise of high interest rates, which he supports, is designed to change the public's expectation that things will cost more next year. There are already some signs that high interest rates are working to slow the economy, notably a fall in housing starts and a levelling-off of consumer loan demand.

Banks are not responsible for high interest rates, because interest rates are determined by the Bank of Canada, he said. Ultimately, though, it is inflation that leads to high interest rates.

6. a) What is the rationale behind Bouey's sure route to lower interest rates?
   b) How would you answer the question posed in the last sentence? (Hint: During this time period Canada and the U.S. developed huge budget deficits.)

7. The first half of this clipping suggests that higher interest rates can lead to lower inflation, but the second half suggests that higher inflation leads to higher interest rates. Explain this paradox.

E. Forecasting

In his eyes the battle is between the rate-lowering effect of the U.S. recession and the high rate of inflation.

That sums up the problem now facing the interest rate forecasters.

1. Explain how these two effects operate on interest rates in different directions.

Even though interest rates will collapse under the onslaught of recession, the marks of tight money are not easily erased, and the recession is likely to be worse for each country than if its battle against inflation had been less potent.

2. Why will interest rates collapse under the onslaught of recession?

There are basically two main camps of thought. In one camp are those who think Canadian and U.S. interest rates will either stabilize or move up because the U.S. economic outlook is improving and will rebound from the paltry 0.7 per cent real growth in the gross national product in the first quarter. Moreover, they argue, growth in money supply has been way above the U.S. Federal Reserve Board's target.

3. Explain the rationale behind the two reasons given for expecting a higher interest rate.

In economic theory, money is a commodity that responds to the law of supply and demand. When the supply rises, the price—the interest rate—should drop.
But the market for money is perverse. When lenders see the money supply increasing, they think. . . .

4. Complete the argument about to be developed here.

Investors who want an edge in predicting the future course of the economy often look closely at the U.S. Treasury's quarterly refinancing of its debt. Usually strong demand for new treasury issues means that interest rates are likely to fall, while poor demand is an indication that interest rates are apt to rise.

5. Explain the logic behind this thinking.

The past few years have been, in effect, a crash course in basic economics for investors and others with a hand in the game. They're much harder to fool now with actions that seem to improve the situation in the short run but make it worse in the long run.

6. Give an example of a policy action that would improve the situation for investors in the short run, but make it worse in the long run.

F. Lowering Interest Rates

This brings us back to the fears of higher interest rates before the market break. These fears are still potent, especially if investors see through the temporary reduction in interest rates made possible by stepping up the rate of creation of the money supply.

1. a) Explain how "stepping up the rate of creation of the money supply" causes a reduction in interest rates.
b) Why will it only be temporary?

He told a news conference the high interest rates will help break the psychology of inflation—where consumers expect costs to keep rising and so keep spending now despite high prices—and make reductions possible.

MacIntosh said he agrees with the Bank of Canada's stand that inflation has to be broken before interest rates can come down.

2. a) Explain how high interest rates will help break the "psychology of inflation."
b) Why must inflation be broken before interest rates can come down?

The key to containing inflation over the long haul is to lower interest rates, he said. Considering the burden high rates create for consumers, businesses and governments, "it seems difficult to justify our central bank's insistence on keeping interest rates at their present level."

3. What response would the central bank give to this argument?

Policies that would lower interest rates are "the single most important thing government could do to contain inflation on a long-term basis in Canada," according to J.L. Biddell, a member of the federal Anti-Inflation Board.

4. Some might argue that the causal relationship here is the reverse—that lowering inflation leads to lower interest rates. Which view do you support, and what is the logic behind it?

"The principal power of the bank to lower interest rates lies in its ability to contribute to a lower rate of inflation and that takes time," he added.

5. Explain the logic of this statement.

He added that the principal misunderstanding about the Bank of Canada's role in the present situation is that it could achieve "more or less immediately a low level of interest rates if it wanted to."

6. Explain why the Bank of Canada cannot achieve "more or less immediately a low level of interest rates if it wanted to."

While Mr. Bouey acknowledged that the bank can influence short-term rates within a limited range, he argued that if the bank were to try to move outside that limit it would not succeed.

7. Why would it "not succeed"?

G. Policy

Although interest rates were for many years the main policy indicator used by most central banks, the experience with severe inflation beginning in the 1970s made it clear they were fickle guides for the task of ensuring that monetary policy was directed toward price stability.	1. Explain how interest rates could be "fickle guides for the task of ensuring that monetary policy was directed toward price stability."
For some months, the Bank of Canada has reduced the attention it pays to the levels of interest rates and has kept a close eye on expansion of the money supply. This policy change has made market interest rates more responsive to the high rate of inflation and to rate increases in the U.S.	2. a) Why would the Bank of Canada have changed its policy to watching the expansion of the money supply instead of watching interest rates? b) Why should this policy change make "market interest rates more responsive to the high rate of inflation and to rate increases in the U.S."?
"One sign of a recession we don't have at present is an excessive build-up in inventories. Nonetheless, some slowing of the rate of growth is necessary to keep Canada in a sustainable expansion and to avoid a recession." Drummond cautioned that an inflation rate creeping up to the five per cent level will trigger a monetary response from the Bank of Canada (i.e. higher interest rates).	3. Why would the Bank of Canada move to raise interest rates if inflation creeps up?
"But I see no reason to think that the central bank is going to relent. Whether they tighten rates further depends on whether there's a reduction in the rate of inflation."	4. Explain the reasoning behind this statement.
It is now evident that loan demand has dropped sufficiently since the beginning of the year for money supply to grow at a rate of less than 10 per cent at current levels of interest rates.	5. This item suggests that the demand for money controls its supply. How can this be?

The idea of investing in debt securities was a gamble on the government's good intentions to fight inflation. The budget makes it clear that the government has taken this task to heart.

It stresses the fact that the deficit will be covered by sales of securities to the general public rather than to the banking system and to the Bank of Canada in particular.

It is a sign that Ottawa has recognized the need to manage its cash requirements within the limitations of a capital market regulated by a well-determined monetary policy.

6.
a) Explain the nature of this gamble.
b) Of what relevance to this gamble is the fact that security sales will be to the public?
c) What is a well-determined monetary policy? What limitations does it impose in this context?

"People are looking to the bond markets for discipline, forgetting that it's very difficult to stop the economy with higher interest rates alone. What stops it is when it runs out of money."

7. Doesn't running out of money cause high interest rates? Explain what is meant here by the statement that "it's very difficult to stop the economy with higher interest rates alone."

II-8 International Influences

Canada has an open economy. It is also smaller than most other countries with developed economies with which it has economic ties. Having these characteristics means that changes in economic activity elsewhere in the world can have a significant impact on the health of the Canadian economy, and international movements of capital (money) can have a strong influence on the character of Canadian money markets. This should already be evident from previous articles.

Impacts on the Canadian economy from the international sector arise whenever Canada's demand for foreign currency differs from her supply of foreign currency. Demand for foreign currency arises when Canadians wish to purchase imports or buy financial assets (such as bonds) in foreign countries. Supply of foreign currency arises when we sell our goods abroad (exports) or when foreigners buy our financial assets (lend us money). The main variables affecting these elements of the supply of and demand for foreign currency are as follows: Exports are determined by the prosperity of foreign countries and the level of the exchange rate (in conjunction with the domestic price level); imports are determined by domestic prosperity along with imports' prices; and capital flows are affected by interest rate differentials, hedging costs, and anticipated movements in the exchange rate.

The difference between exports and imports of goods is called the balance of trade. When the net value of services sold to or bought from foreigners (transportation costs, for example, or interest payments) is added, this account is called the current account. The difference between foreigners' loans to us (capital inflows) and our loans to foreigners (capital outflows) is called the capital account. The sum of the current and capital accounts is called the balance of payments: the difference between our total supply of foreign currency and our total demand for foreign currency. When supply exceeds demand, Canada has a balance of payments surplus; when demand exceeds supply, it has a balance of payments deficit.

The way in which an imbalance in Canada's international payments affects the economy depends on the type of exchange rate system being employed. With a flexible exchange rate system, any imbalance between the supply of and demand for foreign currency is immediately rectified by a "flexing" of the exchange rate, i.e., the "price" of foreign currency. Thus when Canada's supply of foreign currency exceeds its demand, the exchange rate rises, making our imports cheaper and our exports more expensive. This increases imports (increasing the demand for foreign currency), and decreases exports (decreasing the supply of foreign currency), eliminating the excess supply of foreign currency. A similar reaction characterizes a situation of excess demand for foreign currency.

With a fixed exchange rate system, the exchange rate can no longer flex: The government is obliged to exchange currencies at the fixed rate, regardless of whether or not the supply of foreign exchange equals its demand. To see what impact this has on the economy, let us trace through an example of a balance of payments surplus (an excess supply of foreign currency). Those holding the excess supply of foreign currency will ask the government to exchange it at the fixed rate. The government usually buys this foreign currency with money it has received from others to whom they have sold foreign currency, but this cannot be done in this case because more foreign currency is being sold to the government than is being bought from it. Therefore those selling the excess supply of foreign currency will be paid for it with new money; a balance of payments surplus will thus cause an increase in the supply of money. Similarly, a balance of payments deficit causes an automatic decrease in the money supply under a fixed exchange rate system. The only circumstance in which this will not occur is when the monetary authorities undertake an offsetting action, "sterilizing" this change in the money supply.

If the monetary authorities do not conduct sterilization operations, this automatic reaction of the economy (changing the money supply) will cause the imbalance in international payments to disappear. For the example of a balance of payments surplus, there is an automatic increase in the money supply, which has the following effects:

a) Canada's income level is stimulated, increasing imports and thereby increasing the demand for foreign currency;

b) Canada's price level should rise, making imports more attractive and exports more expensive—this should increase the demand for foreign currency and decrease its supply; and

c) Canada's interest rate should fall, changing the difference between Canada's rates and foreign rates. This should decrease net capital inflows (or increase net capital outflows), decreasing the supply of foreign currency (or increasing the demand for foreign currency).

If the monetary authorities do conduct sterilization operations (i.e., buy or sell bonds to offset any change in the money supply), this automatic adjustment mechanism will not operate, and the imbalance in international payments will continue year after year. A continued deficit will eventually cause Canada to run out of foreign exchange reserves, precipitating an exchange rate crisis and probably a devaluation. A continued surplus causes Canada to gradually accumulate a huge quantity of foreign exchange, angering other countries to the point of precipitating an exchange rate crisis and probably an upward revaluation of the Canadian dollar.

The forces described above can be used to explain two very useful theorems, the *purchasing power parity* (PPP) theorem and the *interest rate parity* (IRP) theorem. These theorems provide a background or foundation of *long-run* influences characterizing an economy relative to its trading partners. The PPP theorem in essence says that, after accounting for the exchange rate, (traded) goods and services should cost the same in Canada, say, as in the U.S. In particular, this implies that the difference between the Canadian and U.S. inflation rates should determine the rate at which the exchange rate between their two currencies is changing. Suppose, for example, that the rate of inflation in Canada is 6 per cent and the U.S. is 9 per cent. Then each year, if the exchange rate does not change, Canadian goods become 3 per cent cheaper to Americans and U.S. goods become 3 per cent more expensive to Canadians. As a result, Canadian exports rise and imports fall, creating a balance of payments surplus. Then one of two things happens, depending on the nature of the exchange rate system:

a) Under a fixed exchange rate system, the balance of payments surplus causes Canada's money supply automatically to increase (as described earlier) by an extra 3 per cent, pushing inflation in Canada up to 9 per cent to match the inflation in the U.S.; this eliminates the balance of payments surplus. Notice that Canada is forced to experience the same money supply growth and inflation as the U.S.

b) Under a flexible exchange rate system, the balance of payments surplus pushes up the value of the Canadian dollar by 3 per cent, eliminating the Canadian price advantage caused by the difference in their inflation rates. Notice that this 3 per cent revaluation in the exchange rate is an annual change; as long as the difference between the Canadian and U.S. inflation rates is maintained at 3 per cent, the value of the Canadian dollar must steadily rise at an annual rate of 3 per cent.

The IRP theorem says that the real rates of interest in Canada and the U.S. should be the same. Continuing our numerical example, suppose the real interest rate is 4 per cent, implying that the nominal interest rate in Canada is 4 + 6 = 10 per cent and the nominal interest rate in the U.S. is 4 + 9 = 13 per cent. Suppose that you invest in New York instead of your native Toronto because you can earn an extra 3 per cent on American bonds. When the bond matures at the end of the year, and you attempt to return your money to Toronto, you discover that during the year, as prophesized by the PPP theorem, the value of the American dollar fell by 3 per cent, exactly wiping out the extra 3 per cent interest earned in New York. From an investor's point of view, any difference in countries' nominal interest rates (equal to the difference between their inflation rates) is exactly offset by the anticipated change in the exchange rate. If there is a difference between two countries' real rates of interest, investors will move funds from one country to the other, thereby forcing these real rates to be the same.

From a pragmatic point of view the important implications of these theorems are that:

a) if Canada wishes to have an inflation rate differing from that in the U.S., it must allow its exchange rate to change steadily;

b) if Canada fixes its exchange rate with respect to the U.S. dollar, it must be prepared to have her monetary policy and inflation rate match that of the U.S.; and

c) although Canada's *real* interest rate is basically determined by the U.S. (because of the large relative size of the U.S.), there is no necessity for Canada's *nominal* interest rate to match that of the U.S.

This last point is a source of great confusion in newspaper discussions of Canada's interest rate behavior.

II-8A Dollar-Defence Fund

Federal fund to defend dollar expands

BY ERIC BEAUCHESNE
Canadian Press

OTTAWA

The federal government's dollar-defence fund continued to balloon last month, expanding by $406.5 million (U.S.) to a record $16.6 billion, Finance Department figures show.

The increase in the government's official international reserves apparently reflects efforts by the Bank of Canada to slow the rise in the value of the Canadian dollar.

The dollar hit a seven-year high of 84.67 cents (U.S.) on Jan. 31, up from about 78 cents a year earlier and up from the record low of just under 70 cents in early 1986.

The reserve fund, made up of gold and foreign currencies, is used to stabilize the value of the dollar. All figures are in U.S. dollars because that currency makes up the bulk of the fund.

The Bank of Canada, which handles the reserve fund for the government, sells reserves and buys Canadian dollars when it wants to slow or stop a slide in the value of the currency.

Analysts say a variety of factors have contributed to the rise in the value of the Canadian dollar against its U.S. counterpart. These include the relatively high level of interest rates in Canada, the strength of the Canadian economy, rising prices for many natural resources, especially metals, and the Canada-U.S. free-trade deal.

In recent months, the Canadian dollar has also risen in value against other major foreign currencies, but a strong dollar is a mixed blessing.

Exporters have complained to the government that the current strength of the dollar is hurting them.

The central bank could reduce interest rates to stop the rise in the value of the dollar or reduce its value. However, the Bank of Canada is keeping interest rates up to combat inflationary pressures, which it argues are the major threat to the economy.

---

1. Explain how "efforts by the Bank of Canada to slow the rise in the value of the Canadian dollar" (second paragraph) would cause the government's dollar-defence fund to balloon.

2. Explain how each of the four factors cited in the sixth paragraph would contribute to the rise in the Canadian dollar. Which of these factors is often associated with downward pressure on the Canadian dollar? Explain.

3. Why would exporters be complaining, as noted in the second-last paragraph?

4. Why would a reduction of interest rates stop the rise in the value of the dollar, as noted in the last paragraph?

5. The central bank's policy of keeping the interest rate high implies that the money supply is being restricted. What central bank activity reported in this article serves to *increase* the money supply? Explain.

II-8B Big Mac Index

Big Maconomics

It takes sophisticated experts in banking and economics to understand the intricacies of world currency values, and none of them has ever been able to explain them satisfactorily to the layman. But everyone can understand the Big Mac Index.

The Big Mac Index is a rudimentary guide to exchange rates based on the price of the hamburger at McDonald's restaurants in countries around the world.

To understand the Big Mac Index you don't need to know anything about merchandise trade deficits, balances on current account, central bank interest rates, money supply, and all that stuff. All you need to know is that if you pay twice as much for a hamburger in one country as you pay in another for the equivalent amount of money, you're being jobbed.

For instance, the latest version of the index shows that in United States dollars, the cost of a Big Mac ranges from 74 cents in Hungary to $2.19 in the U.S. to $4.21 in Norway. Extrapolation of those figures suggests that in relation to the U.S. dollar, the Norwegian kroner is overvalued by 92 per cent, while the Hungarian forint is undervalued by 66 per cent.

By the same measurement, the currencies that the experts get all worked up about, the West German mark and the Japanese yen, are overvalued by two and 27 per cent respectively. (In Vancouver, where a Big Mac costs $1.63 US, the hamburger index shows the Canadian dollar undervalued by 25 per cent.)

Even serious economic publications like *The Wall Street Journal* and *The Economist* have lately started to pay attention to the Big Mac Index, which the *Journal* says was suggested four years ago by Georg Grimm, an adviser to West Germany's Chancellor Helmut Kohl.

"The Big Mac Index is primitive, but anyone can understand it," the *Journal* quotes Herr Grimm. "You can explain it to a child, or to a prime minister."

The difference is that, faced with an overpriced hamburger, the child would probably know what to do about it.

Note: Suggested answers to the questions below appear in the Appendix.

1. What economic theorem is implicitly being discussed (spoofed?) in this editorial?
2. What does this theorem suggest should be the value of the Big Mac Index in countries around the world?
3. Explain the economic forces that are supposed to bring about this result.
4. The facts cited in this article do not support this theorem. Offer some explanations of this.
5. Explain the arithmetic behind the statement that the Canadian dollar is undervalued by 25 per cent.
6. Interpret the message of the final sentence.

Monetary policy 'could be made-in-Canada'

**Finance Letter/
By Barry Critchley**

TORONTO

IT IS A THEME that occupies considerable interest in Canada: does the Bank of Canada have the ability to operate a monetary policy independent of that in the U.S.? And assuming it does, *should* it act independently of U.S. monetary policy?

Not an easy issue when market understanding and public perception is that Canadian rates are determined largely by U.S. events, when the C$ is near its all-time low, when the unemployment rate is high, and when so many politicians, business leaders and union officials say the Bank should act independently—and decisively. The real answer—as indicated in a recent talk by Jack Carr, professor of economics at the University of Toronto—is dependent on the exchange rate regime. If the exchange rate is fixed, we lose our independence and have to follow the U.S. to maintain the peg. If exchange rates are floating freely, Carr says that, contrary to the popular view, there are no limits on the Bank's ability to act independently.

Thus he rejects the view that because we're a small open economy with few restrictions on capital flows, Canadian rates have to be similar to those in the U.S.

The key, Carr says, is what measure of interest rates—nominal or real—is being considered. "Nominal rates in Canada can be independent of U.S. interest rates and the Bank can, if it so desires, follow a monetary policy independent of the U.S. situation." Thus, if U.S. authorities decide to expand the money supply—which, in the long run won't change real interest rates, just the nominal rates—Carr says we don't have to follow in lock-step.

In such a situation, Carr argues that the appropriate Canadian response would be to ignore U.S. developments and maintain policy, which would be geared to producing a lower-than-U.S. inflation rate.

In turn—and assuming perfect information by market participants of this change—there wouldn't be any capital outflow and there wouldn't be any pressure on the exchange rate. In fact, the C$ would appreciate and everything else being equal, that appreciation would equal the difference in the inflation rate. "From a theoretical point of view, Canadian and U.S. inflation rates don't have to be equal."

While this monetarist interpretation shows there is full independence for the setting of nominal interest rates—the Bank determines the inflation rate—Carr says the Bank doesn't have any independence when it comes to real interest rates.

'Nothing we can do'

"There is nothing we can do about it even in the short run, because the real rate is determined in international capital markets," he says, adding that under floating rates, and in the absence of capital restrictions and taxation differences, real rates around the world are equal over time.

So, what should the Bank have done during the most recent runup in U.S. interest rates? Unlike most economists—who argue that because of the speculative pressures on the US$, the most appropriate action was to let the C$ slip—Carr says the Bank should have maintained its previous thrust of a reasonably moderate monetary policy. This would have meant continued low inflation and, in time, interest rates lower than in the U.S.

In fact, though, because of its concern for the short run, the Bank engaged in a moderate expansionary monetary policy "designed to keep interest rates low."

Even with full independence, Carr doubts whether we would be much better off because of it—in part because of political pressure acting on the Bank. That pressure is to continuously inflate and blame monetary policy for a multitude of problems—even if the solution lies with other policies. Another negative: a made-in-Canada monetary policy won't produce a made-in-Canada real interest rate.

---

1. Explain the difference between the nominal and real interest rates.
2. Explain why an expansion of the U.S. money supply will raise the nominal interest rate, as noted in the last sentence of the fourth paragraph.

3. In the first sentence of the sixth paragraph the author says that "there wouldn't be any pressure on the exchange rate" but then in the next sentence states that "the Canadian dollar would appreciate." Explain what he's really trying to say here.

4. Explain why there wouldn't be any capital outflow, as stated in the first sentence of the sixth paragraph.

5. Explain how capital restrictions and taxation differences (fourth-last paragraph) can cause real rates to differ between countries.

6. Explain in detail why "If the exchange rate is fixed, we lose our independence." (second paragraph)

Short Clippings

Note: Suggested answers to the odd-numbered questions below appear in the Appendix.

A. **Trade Deficits and the Current Account**

Vasic says an overvalued C$ would be just another blow to an already fragile economy. Consumers, the main engine of economic growth, are already over-extended and the economy is still feeling the aftershocks of the stock market crash, he says.	1. Use an aggregate supply-aggregate demand diagram to illustrate how "an overvalued C$ would be just another blow to an already fragile economy."
The rise of the C$ vs the US$ could batter the profits of some Canadian companies this year and slice into capital spending.	2. a) Why would a rise of the C$ "batter the profits of some Canadian companies"? b) Why would it slice into capital spending?
When assessing the effect of the U.S. growth slowdown on Canadian exports, it is important to bear in mind that much of the U.S. slowdown is due to the diversion of U.S. demand abroad.	3. a) What would cause a diversion of U.S. demand abroad? b) Why would this cause a U.S. growth slowdown?
In a recent analysis, Wood Gundy Ltd. of Toronto concluded that in many export sectors the devaluation of the dollar will not greatly aid sales. Progress will be slow, the investment firm stated, because any U.S. slowdown would "overshadow the price effects," of the devalued dollar.	4. What is meant by a U.S. slowdown overshadowing the price effects of the devalued dollar?
A major negative influence on the U.S. economy has been the strong dollar. Its strength causes excessive imports. The excess of imports over exports—more than $120 billion (U.S.) a year—must be subtracted when calculating the GNP.	5. a) Why is the strong U.S. dollar a negative influence? b) What method of calculating GNP requires that the excess of imports over exports be subtracted? From what is it subtracted? Why is it subtracted?
Looking at the data, some people might say the Canadian balance of payments has "improved" radically, from a $5.8 billion current account deficit in 1981 to surpluses of $3 billion and $1.6 billion in 1982 and 1983, respectively. It is this ongoing current account surplus which led the optimists to predict that the C$ should appreciate.	6. What error is made by those who look at this data and say that the Canadian balance of payments has improved?

The link between the current and capital accounts is often misunderstood. Leaving aside the relatively small influence of central-bank intervention in foreign exchange markets, any deficit or suplus on capital account *must* be matched by an equal and opposite surplus or deficit on current account. It follows that, as long as the U.S. is to be a capital importer, it *must* have. . . .	7. a) Complete the last sentence. b) Explain why the matching described here must occur.
The U.S. trade deficit is not permanent. When the U.S. ceases to be a major importer of capital. . . .	8. Complete this argument for the reason the U.S. trade deficit will disappear.
While some in the U.S. put the trade deficit down to failing U.S. competitiveness or protectionist policies abroad, economist Richard Lipsey, of Queen's University and the C.D. Howe Institute, said its genesis lies in the budget deficit, and the consequent shortfall in U.S. domestic savings relative to investment.	9. Explain the logic behind how the U.S. budget deficit could be responsible for the U.S. trade deficit.
The Canadian dollar is now worth about 77.7 per cent of the U.S. unit. A year-and-a-half ago it was 71.9 per cent. That is a recipe for a serious deterioration in our current account position.	10. Do you agree with this conclusion? Explain why or why not.

B. Determinants of the Exchange Rate

The Reichmann family's $3 billion acquisition of Gulf Canada Ltd. from Chevron Corp. of San Francisco and summer seasonal weakness will keep the Canadian dollar under pressure.	1. What kind of pressure will the Canadian dollar be under? Explain.
Some economists also say a downturn in commodity prices could hit the C$ hard.	2. What is the rationale behind this claim?
The Canadian dollar rose from an all-time low of 69 cents (U.S.) in February, 1986, to more than 81 cents in April. Finance Minister Michael Wilson has repeatedly said the main reason behind the rise in the dollar has been the strength of the Canadian economy.	3. a) What is the usual impact of a strengthening economy on the exchange rate? Explain. b) How can Wilson's claim be defended?

Fowler says a firming in commodity markets; a shortage of good investment prospects for the huge amount of international money in the system; lower U.S. interest rates; and a falling US$ are behind the strong C$.

4. Explain how each of these factors would serve to strengthen the C$.

"Everyone loves the Canadian dollar these days and the fall in the current account deficit just added to its popularity," said Mike Bowick, a currency trader at the Royal Bank.

5. Why would a fall in the current account add to the "popularity" of the Canadian dollar?

That's what has Canadian officials worried: "The floating exchange-rate system has served us well," one says. Adds a recent Ontario government study: "In recent years, Canadian currency depreciation has facilitated adjustment, first to higher inflation, and then to sharp declines in commodity prices and sales volumes."

6. Explain how Canadian currency depreciation has "facilitated adjustment" to higher inflation and declines in commodity prices and sales volumes.

In the case of the Canada-U.S. exchange rate, the statistical evidence (when changes in relative price levels are controlled for) is that there is a strong correlation between increases in the current account surpluses and declines in the value of the C$. This is just the opposite of the conventional wisdom.

7. a) Explain what logic lies behind the "conventional wisdom."
b) How would you explain the statistical facts cited?
c) Why are changes in relative price levels controlled for?

The report concludes that the recent strength in U.S. investment has been financed by domestic investors redirecting funds from foreign capital markets to U.S. capital markets. Thus, increased demand for US$s isn't the reason for the strength of the US$, but rather a decreased U.S. demand for foreign currencies.

8. What would have caused U.S. investors to redirect their funds?

C. Fixing the Exchange Rate

Furthermore, current monetary policy appears to be conducted as if we were on a fixed exchange rate. Policy is effectively geared to maintaining the exchange rate at the expense of a domestic recovery.

1. Explain how a policy geared to maintaining the exchange rate would prevent a domestic recovery.

A second issue concerns how long the fixed parity could be maintained. This, of course, depends on an uncertain future, but if the initial value chosen was appropriate and if domestic costs remained in line with foreign costs, then the rate may be sustainable for some time. For example, the parity established in 1962 lasted for eight years.

2. a) What is meant by "if domestic costs remained in line with foreign costs"?
   b) What would happen if the government was determined to keep the exchange rate fixed but domestic costs became greater than foreign costs?

Finally, there is the question of monetary policy and imported inflation. Under a fixed exchange rate, Canada's inflation would be more closely tied to that of the U.S. Given the experience of the past few years, that may not be a bad thing.

3. a) Explain the rationale behind the second sentence.
   b) On the basis of the comment offered in the last sentence, what change in current monetary policy would come about if the exchange rate were fixed?

If Canada adopted a system of fixed exchange rates, Canadian currency would have to be exchanged at fixed rates with the U.S. currency. Under such a system the monetary policy in Canada must be identical to the monetary policy followed in the U.S. Under fixed exchange rates, the role of the Bank of Canada is severely limited.

4. Explain why it is that monetary policy in Canada must follow monetary policy in the U.S. if there is a fixed exchange rate.

Added to the bank's potential problems is concern that the U.S. might not be as dedicated as Canada is to fighting inflation. Asks Beigie: "What if inflation rises significantly in the U.S.? The bank hasn't considered this possibility."

5. Explain clearly the problem to which Beigie is referring.

Individual nations do not want to sacrifice their independent monetary policies, but experience has shown that the only route to stable exchange rates is through similar monetary policy. If Canada is to take a lead role in this drive for stability then we must reach a consensus from the beginning that the idea of a "made in Canada" monetary policy is not realistic.

6. a) Explain the rationale behind this thinking.
   b) How could a "made in Canada" monetary policy be achieved?

This is the reason why the fixed exchange rate system was scrapped in 1971. The U.S. had been pursuing an inflationary monetary policy to help pay for the Vietnam war and new social programs, and its trading partners did not all want to participate in it.

7. Explain the rationale behind this statement.

To be sure, Crow said the broad money aggregates M2 (currency plus chequeable and nonchequeable bank deposits) and M2+ (all of the above plus the same at trust companies and credit unions) have merely been upgraded to "indicative policy guides," rather than formal targets. And while he made plain his rejection of the former unspoken policy of targeting the US$ exchange rate, he said the Bank would still keep one eye on the C$, against a trade-weighted basket of major currencies. But the message was clear: to read the future of Bank policy, check the M2 numbers. It has been many years since the Bank has been quite so explicit.	8. Explain why adoption of this new policy would require that the former policy be rejected.
The Bank of Canada, since early 1985, has all but pegged the C$ at around 73¢ U.S., give or take a percent or two (and leaving aside the crisis weeks of early last year). To do this over any extended period of time requires that our monetary growth rate be attuned to that of the U.S.; Canadian M2 growth has, indeed, also averaged 9 per cent or so. With the exchange rate as the Bank's monetary north star, Fed policy has been our policy.	9. Explain the meaning of and rationale behind the last sentence.

D. Managing the Exchange Rate

The Bank of Canada, which handles the reserve fund for the Government, sells reserves and buys Canadian dollars when it wants to. . . .	1. Complete this clipping.
The central bank, fearing inflation, has kept interest rates up. However, to offset the upward pressure those rates are putting on the currency, it has also been. . . .	2. Complete this clipping.
The government dumped close to $3 billion US worth of Canadian dollars on world money markets to. . . .	3. Complete this clipping.

When Ottawa wants to strengthen the dollar it _____ foreign reserves and _____ Canadian dollars. It does the opposite when it wants to keep the dollar from rising too sharply.	4. Fill in the blanks in this clipping.
By keeping interest rates high, the bank has attracted offshore investment and created a demand for Canadian dollars. In turn, this has pushed up the dollar, taking millions of dollars off the bottom lines of exporting companies. To curb the dollar's climb, the bank has periodically entered the money markets. . . .	5. a) Explain how millions of dollars were taken off the bottom lines of exporting companies. b) Complete the final sentence by specifying the nature of the bank's activity on the money markets.
Canada's international reserves fell by $314.7 million to $3.1 billion last month—not much ammunition if the going gets tough.	6. a) What must have been happening to cause the reserves to fall? b) What does "not much ammunition if the going gets tough" mean?
It is surprising that, with the dollar trading above 86 cents (U.S.), the Bank of Canada has not allowed some easing of interest rates at the expense of a weaker dollar. "Since North America has entered into recession and exports are falling off, clearly the capacity constraints argument with respect to export industries carries less weight."	7. What is the "capacity constraints argument" and why is it relevant here?
The two-pronged attack—raising interest rates to attract liquid capital into Canada and using foreign currency holdings to sop up unwanted Canadian dollars—has been designed to cushion the fall of the domestic currency, which has dropped by about three cents in the past two months.	8. a) Explain how these two actions cushion the fall of the Canadian dollar. b) Which is only of short-term validity? Why?
Then there's the state of the dollar, which has been bleeding steadily despite transfusions from borrowings and foreign reserves. Will the higher bank rate stop the hemorrhage?	9. a) What does the dollar "bleeding steadily" mean? b) Explain exactly what "transfusions from borrowings and foreign reserves" are and how they would help stop the bleeding. c) How would a higher bank rate "stop the hemorrhage"?

When the buck did collapse, with capital flowing out after the Parti Quebecois victory, Dobson says, the Bank resisted the depreciation—which would drive up import prices and, therefore, domestic inflation—by buying up unwanted C$ in the market, by placing federal debt offshore, and by linking Canadian interest rates to U.S. rates.

10. Explain how each of the three methods referred to can act to resist depreciation.

The government, together with the Bank of Canada, can influence the rate through market intervention or shifts in monetary policy. However, since the C$ was allowed to float freely six years ago, Ottawa has practised a hands-off policy except to smooth day-to-day fluctuations in the market. The view is shared by foreign observers, some of whom might be only too delighted to be able to claim that Canada is a dirty floater.

11. a) Explain how the actions cited in the first sentence can influence the exchange rate.
    b) What is a "dirty floater"? Why might a country want to be one?

Ottawa might be tempted to increase official currency reserves in order to reduce the pressure on the C$ and the consequent threat to manufacturing industries. But such a move would be difficult. The government would need to borrow large sums on the market because its cash balances are not too high, considering that cash requirements will be close to $6 billion again this year.

12. a) How would increasing official currency reserves affect the exchange rate?
    b) How would this reduce the "threat to manufacturing industries"?
    c) Why would increasing official currency reserves be difficult? Can't the government just print more money?

When the Bank of Canada's foreign exchange traders go into the market to support the dollar, they trade United States dollars from Canada's reserves for Canadian dollars other people are trying to sell.

The Canadian cash resulting from this transaction goes into the government's ordinary account, reducing Mr. Chretien's need to borrow from other sources to cover the deficit. Since the beginning of the fiscal year, about $2 billion in Canadian funds has appeared from this source.

13. a) Was the Canadian dollar trying to rise or to fall during the period referred to in this clipping? Explain your answer.
    b) If this is such a good way of supplementing tax revenues, why not do more of it, more often?

The response, as Bouey relates it in his report, has been to search out the middle ground. The Bank has a simple operating formula: Some of the adjustment will be taken through the exchange rate, some through interest rates, and the rest through a loss of international reserves.

14. To what is the Bank adjusting here? Explain how these mechanisms work and in what direction they adjust.

II-8 *International Influences* 219

These are the extreme positions. The Bank's actual path is somewhere in the middle. Some of the brunt of higher U.S. interest rates is taken in higher domestic rates, some through a lower-valued C$ and some through a loss of international reserves.

15. Explain how a loss of international reserves plays a role here.

Official intervention (in the foreign exchange market) does appear to be a useful policy, when overshoots take place in fragile circumstances, he said.

16. What is official intervention? Exactly what would it involve in this case of an overshoot?

The minutes also reveal that stemming the rise in the value of the dollar forced the government to borrow heavily on domestic financial markets.

17. Why would the government have to "borrow heavily on domestic financial markets" to stem the rise in the value of the dollar?

Reagan's own Council of Economic Advisers has put the case for a muscular dollar with surprising force.
"The strong dollar has stimulated production and investment in sectors less involved in international trade," the council's recently issued 1985 annual report to Congress states. "In other industries, competition from imports has prompted more expenditure in plant and equipment as well as greater attention to controlling wages and other costs. Prices of traded goods and close substitutes have been kept lower than they would have been otherwise, thereby benefiting both U.S. consumers and U.S. producers who use imported inputs."

18. a) What is the basic argument against a muscular dollar?
b) In a half-dozen words, summarize the case supporting a muscular dollar.

E. Purchasing Power Parity

Floyd says that changes in the currency's value are a safety valve. Over the long run, it will move to match up the changes in domestic prices with those prevailing abroad.

1. What technical terminology do economists employ to describe this phenomenon?

Hence, to the extent that the decline of the foreign currency relative to the C$ matches the differences in the two countries' inflation rates, the traveler probably isn't better off. The extra foreign currency will be required just to pay for the higher-priced goods and services.

2. What economic theorem does this clipping describe?

The exchange rate between two countries usually bears some relationship to the difference in inflation rates between those same countries. Known formally by economists as purchasing power parity, that theory asserts that over time, the change in the exchange rate will match the change in the inflation rate.

3. This author's version of purchasing power parity is carelessly worded. Remove the ambiguities.

While the table shows that there are more countries against which the C$ has appreciated over the past 12 months rather than depreciated, the extra local currency that a fixed C$ outlay will buy is not necessarily a great windfall.

4. Explain why this might be the case.

While critics have suggested that the bank could lower rates and, in so doing, let the dollar fall on international exchange markets, the report said that the dollar has already fallen in the last half-dozen years and raises the question of why things have not gone better in Canada if exchange rate depreciation is so good for the economy.

5. Offer an explanation for "why things have not gone better in Canada" as a result of the dollar depreciation.

The Department of Labor figures are based on so-called "purchasing-power-parity exchange rates." The difference in Canada's case is substantial. For example, if the official C$/US$ exchange rate was used in 1984, our gross domestic product per capita would have been only US$13,200 or 14 per cent below the U.S. figure instead of only 4 per cent lower.

The official exchange rate doesn't capture the fact that a US$, exchanged into C$, can actually buy more goods and services in Canada than it can south of the border. The purchasing-power-parity exchange rate adjusts for real differences in purchasing power.

6. a) Explain how you would calculate the purchasing-power-parity exchange rate.
b) Why would a difference arise between the PPP rate and the regular exchange rate?
c) In what direction was this difference in 1984?

Canadian living standards are second only to those in the U.S., according to the Organization for Economic Co-operation & Development. Using a measure called purchasing power parity—which tries to standardize the buying power of various currencies—the OECD estimates that per-capita gross domestic product in Canada last year was US$15,700. This is 91 per cent of the U.S. standard of living, which is US$17,200. Norway (US$15,100) and Luxembourg (US$14,300) were the only other nations that approached the Canadian living standards. These were followed by Sweden and West Germany with roughly US$13,000 each.

When all currencies are simply converted to a common measure, however, six countries besides the U.S. surpass Canada in nominal per-capita wealth—Norway, Japan, Sweden, Denmark, West Germany, and Finland.

Big Mac-watchers will rely on the theory of purchasing power parity (PPP) for currencies. This argues that an exchange rate between two currencies is in equilibrium (at PPP) when it equates the prices of a basket of goods and services in both countries—or, in this case, that rate of exchange which leaves hamburgers costing the same in each country. Comparing actual exchange rates with PPPs is one indication of whether a currency is under- or over-valued.

So even though PPPs are handy for converting living standards into a common currency, they are not necessarily the best way to judge the exchange rate needed to bring the current account of the balance of payments into "equilibrium."

7. Explain exactly how the calculation undertaken to obtain the results reported in the first paragraph differs from that used to obtain the results reported in the second paragraph.

8. a) Explain, by means of an explicit example, how you would use the price of hamburgers to find the PPP exchange rate to thereby determine whether a country's currency is "under- or over-valued."
   b) What is the major objection to using the hamburger as a basis for calculating the PPP exchange rate?

9. Suppose Canada's exchange rate were "the exchange rate needed to bring the current account of the balance of payments into "equilibrium."
   a) Would Canada's balance of payments be in balance, in deficit, or in surplus?
   b) Does this suggest that Canada's PPP exchange rate is above or below Canada's actual exchange rate? Explain your reasoning. Hint: Recall that Canada usually experiences a net capital inflow.

F. Inflation and the Exchange Rate

Last week, Bank of Canada Governor Gerald K. Bouey said in a speech in Fredericton: "There is the possibility, in theory at least, that a continuing higher rate of inflation in Canada than in the U.S. could be accommodated by a movement over time of the exchange rate. But it would be a serious error to suppose that such accommodations would work smoothly. One has only to look at the experience of other countries to see how disruptive the movements in exchange rates can be between countries with appreciably different patterns of inflation."

1. Explain how "a continuing higher rate of inflation in Canada than in the U.S. could be accommodated by a movement over time of the exchange rate." Would the exchange rate move up or down?

In three to five years, the U.S. dollar will presumably resume its long-term slide unless Washington reverses its economic policies of the post-Second World War period and takes a tough stand against inflation. Observers believe a Reagan administration may take that tougher stand.

2. Explain the logic behind the first sentence.

The conventional wisdom has been that inflation is bad for the economy, that it costs jobs and cuts into growth.
    If Canadian inflation is running higher than American inflation, according to this argument, Canada's exports become non-competitive, growth slows and jobs are lost.

3. a) Explain the mechanism whereby the results forecast in the second paragraph would be brought about.
   b) Explain how this problem could easily be avoided.

This is so even if our goal is not the external value but rather the domestic value of the Canadian dollar, and even if the surest way to sustain the exchange value of our currency is to maintain its purchasing power at home.

4. Explain why "the surest way to sustain the exchange value of our currency is to maintain its purchasing power at home."

Monday's trade figures—twice as bad as analysts were generally expecting them to be—reinforced fears that the British economy was overheating, with imports outstripping exports and inflationary pressures surging.
    The government is now expected to raise interest rates almost immediately.

5. a) Would you expect imports to outstrip exports in the presence of inflation? Explain why or why not.
   b) Why would it be expected that the government would "raise interest rates almost immediately"?

"John Crow (Bank of Canada governor) has done an incredible job convincing the world he's an anti-inflationist," which has helped the Canadian dollar, said Sherry Atkinson, chief economist with Burns Fry Ltd.

6. Explain why Crow's "incredible job" has "helped" the Canadian dollar.

News that U.S. job creation in January was more robust than anticipated sent a signal to currency markets to expect a stepped-up fight against inflation, unleashing a bout of buying fervor for the U.S. dollar.

7. a) Why would robust job creation lead to expectations of a "stepped-up fight against inflation"?
b) Why would this in turn unleash "a bout of buying fervor"?

Word that growth in the cost of living was slower in Canada than in the United States last year attracted buyers for the currency, after a quiet period last week as the market awaited the November trade results for both countries.

8. Explain the logic of this.

In Charlottetown, Mr. Bouey took a different approach. While acknowledging that a weaker dollar initially favors exports, he said this advantage will only last as long as. . . .

9. Complete Bouey's argument.

"The record has not been good," he told the board of directors. "Our past experience with exchange rate depreciation has been that all too often it has led, not to a sustained improvement in our competitiveness, but to. . . ."

10. a) How would exchange rate depreciation lead to an improvement in our competitiveness?
b) Complete Bouey's statement.

G. Interest Rate Parity

At the same time that Secretary Blumenthal was testifying to Congress, the Treasury borrowed $1.6 billion in Germany in the form of securities denominated in marks. It offered to pay an interest rate of roughly 6 per cent per year on mark-denominated three- and four-year securities. On comparable securities denominated in dollars, the Treasury is currently paying a bit over 9 per cent—or 3 percentage points per year more.

1. a) Why is the rate of interest offered on mark-denominated securities less than that offered on dollar-denominated securities?
b) What does the U.S. Treasury believe about the future value of the U.S. dollar in terms of marks? What do German investors buying these securities believe? Explain your answer.

Hart can't understand why Canadians would put their money in three-year paper at 9 per cent, when they can get double-A rated New Zealand bonds at 19 per cent.	2. What explanation would you offer for this?
Before the advent of modern international financial markets in the mid-1970s, domestic financial markets placed a serious constraint on the ability of governments to borrow their way to popularity or prosperity. Historically, domestic financial markets were largely closed systems where excessive borrowing by governments led quickly to. . . .	3. a) Complete this clipping. b) How has "the advent of modern international financial markets" changed this?
Higher interest rates, therefore, are inappropriate for Canada. But in view of Canada's close and direct economic and financial links with the United States—closer than that of any other country—Canada is not free to pursue an interest rate policy more suited to the country's economic position. Mr. Bouey is convinced that adoption of a made-in-Canada interest rate policy would expose the country to the risk of. . . .	4. Complete Gerald Bouey's statement.
While the chartered banks have so far refrained from raising their prime rates, Mr. Lalonde warned reporters yesterday that this situation cannot last. The Canadian and U.S. economies are so interrelated, he said, that it will be impossible for domestic interest rates to remain much lower than U.S. rates for an extended period of time.	5. What qualification needs to be added to this statement to make it valid?
If Reagan doesn't bring rates down, a crumbling U.S. economy will.	6. Explain how a crumbling U.S. economy would bring down U.S. interest rates. How would this in turn bring down Canadian interest rates?
The governor of the Bank of Canada today urged Canadians to continue the fight against inflation, but warned they cannot expect any significant drop in interest rates until the U.S. takes convincing action to reduce its deficit.	7. Explain why this statement could be viewed as misleading.

II-8 *International Influences*

U.S. money supply in the October-to-March period grew at a 10 per cent annual rate, compared with growth of less than 1 per cent between June and October last year.

This behavior tends to strengthen other currencies, including the Canadian dollar, as it becomes progressively less profitable to invest in U.S. dollar-denominated, short-term investments.

8. Explain the logic of this argument.

This does not mean there is a massive flow of U.S. funds into Canada, because the differentials are mitigated by other factors. In the case of market rates, the discount on the forward C$ has kept step with the interest rate differential, and it is only on an unhedged basis that the full advantage of the differential can be gained.

9. Explain the meaning of the second sentence.

H. Interest Rates and the Exchange Rate

Exporters and others have complained that the strong dollar threatens their international competitiveness and have urged the Bank of Canada to lower interest rates, which they say are unnecessarily high and have helped push the dollar to its current level.

1. a) Explain how the strong dollar threatens the international competitiveness of exporters.
b) Explain how high interest rates would "have helped push the dollar to its current level."

Strange things are happening in Canadian financial markets this festive season. In a rare display of independence, our interest rates have bucked the downward trend of rates south of the border. And the dollar, which usually has the blahs at this time of year, is putting on a show of strength.

2. Is the "show of strength" by the Canadian dollar what one would expect given that "our interest rates have bucked the downward trend of rates south of the border"? Explain why or why not.

Economists and important figures in the Reagan administration have been debating whether it is government spending deficits, restrictive monetary policy, capital inflows, or anticipated inflation that is the real culprit for interest-rate increases in the U.S.

3. a) Which of these four possible reasons could definitely *not* have caused high interest rates? Why?
b) Explain how the other three could have led to high interest rates.

The other main method of intervention is to raise interest rates. The Bank of Canada's lending rate went to 12.76 per cent Thursday, and the chartered banks immediately began raising the prime rates they charge on loans to their best-heeled corporate customers.

The rise in rates is more serious than the fall in reserves because it can make the fall in the dollar self-reinforcing instead of self-righting. Higher rates will slow down the Canadian economy even more, producing a bigger growth gap compared to the U.S. economy.

4. a) The first main method of intervention presumably had something to do with "the fall in reserves." Explain what this intervention must have been.
   b) Rebut the argument presented in the second paragraph.

As the dollar continues to fall and Canadian interest rates are held below their U.S. counterparts, some observers detect a subtle change in policy by the Bank of Canada. Despite its repeated denials, it does appear to be willing to accept a lower dollar rather than cripple the economy with. . . .

5. Complete the last sentence of this clipping and explain the rationale behind your choice.

The economic recovery abroad is the most likely key to a stronger C$, the stabilizing of interest rates, and the possibility of lower interest rates here.

6. Explain how the economic recovery abroad could be the key to lower interest rates in Canada.

What would happen, for example, if Michael Wilson adopted a deliberate policy of deficit reduction and lowering interest rates? Listen to the market through the voice of Richard Kapsche, vice president and futures floor manager for E.F. Hutton at Chicago's International Monetary Market:

"We would read that as an inflationary policy and start selling off Canadian dollars."

7. a) Would this policy be inflationary? Why or why not?
   b) Why would the market start selling off Canadian dollars if it perceived that the Canadian government was adopting an inflationary policy?

Because Canadian inflation will continue to accelerate while inflation in the United States starts to decline, the Canadian dollar will probably come under pressure in 1981. This will force the Bank of Canada to keep interest rates up, and "with inflation accelerating, you have to be very careful about bonds."

8. a) Explain the logic behind believing the Canadian dollar "will probably come under pressure in 1981."
   b) Why will this force the Bank of Canada to keep interest rates up? Is this an appropriate move on the part of the Bank of Canada? Explain why or why not.
   c) Why does one have to be very careful about bonds when inflation is accelerating?

Few people are recommending a devaluation of the dollar. Rather, they are saying "lower interest rates and accept the falling dollar that follows." Bouey probably has a two-part response to that, the first being that he can't lower interest rates since they are set by the market. That remains to be seen. He can certainly add money to the system (through purchasing bonds, for example) and that will drive rates down unless inflation fears go up.

9. a) Why will the dollar fall if interest rates are lowered?
b) Why will adding money to the system drive interest rates down?
c) Why might inflation fears go up?
d) What is your opinion on the probable success of the action suggested in the last sentence? Explain your reasoning.

The Bank has made occasional attempts to narrow the gap between the two sets of rates and accept some lowering of the C$ as a tradeoff. Sometimes it works, and sometimes, like last summer, we end up with the worst of all possible worlds—higher interest rates and a lower C$.

10. What would cause this "worst of all possible worlds" to occur?

Turning our currency around requires convincing investors, speculators and the Canadian public at large that the C$ will hold its value. For the moment, that means we must decisively match U.S. interest rates. Paradoxically, in the past we have usually ended up with the worst of both worlds: higher interest rates and a devalued dollar.

11. a) Construct a scenario in which this recommended policy of matching interest rates could lead to exactly the paradox cited.
b) Why would the author have called it a paradox?

The April U.S. trade deficit narrowed by 41 per cent from March. Investors greeted the news with gusto, driving bonds more than a point higher in minutes.

It was a classic suckers' rally. Traders soon realized the trade improvement was because of a 23 per cent climb in exports. Fears of surging demand in an environment of near-full employment and capacity constraints were rekindled. Prices turned on a dime and wound up nearly two points off their highs by the end of the day.

12. a) Explain why the narrowing of the trade deficit would cause the price of bonds to rise.
b) Explain in your own words why bond prices then fell, as reported in the second paragraph.

I. Policy

This reality makes fiscal policy operations, dollar for dollar, far more effective in the United States than in Canada and the OECD study bears this out. In essence, the fiscal policy multipliers in the United States for identical tax or expenditure devices are larger than their Canadian counterparts.	1. Explain why the relative openness of the Canadian economy makes these multipliers smaller.
The Bank of Canada isn't ready to lay down its main weapon—high interest rates—in the fight against inflation despite signs that prices have stabilized and that exporters are being hurt by central bank policy.	2. Explain why exporters would be hurt by this policy.
Last week's movement was largely triggered by Bank of Canada concerns about the drop in the C$ in the face of higher U.S. interest rates.	3. What is the "movement" referred to in this clipping?
If inflation exists in the rest of the world but not in Canada, our merchandise trade surplus should soar with exports becoming cheap and imports expensive. Of course, that would stimulate economic activity in Canada and, pretty soon, the inflation rate would begin to rise. How would Crow counter that undesirable development? By keeping money growth low and interest rates much higher than U.S. interest rates—the same way that the Bank of Canada has fought inflation in the past and still does today.	4. a) What exchange rate system must this author be assuming? Explain your answer. b) Comment on the suitability of the monetary policy that this author claims Crow would employ.
Why has Mr. Crow argued for a stable Canadian dollar and zero inflation? Aren't these inconsistent goals?	5. Explain why these goals could be inconsistent.
Many economists question whether it's possible to achieve zero inflation and a stable currency, as Crow is attempting. Says Carl Beigie, chief economist at McLean McCarthy Ltd. in Toronto: "It seems to be an overdetermined policy."	6. Explain why this would seem to be "an overdetermined policy."

By avoiding policies to slow the growth of domestic demand and instead forcing the U.S. dollar to serve as the adjustment vehicle in narrowing the trade deficit, the United States will ensure that the global economic landscape during the next five years will be turned upside down.

7. a) How would slowing the growth of domestic demand narrow the trade deficit?
   b) What is meant by "forcing the U.S. dollar to serve as the adjustment vehicle"?

With a considerable amount of slack in the economy (following the recession of 1981-82), the main risk on the inflation front was from price shocks caused by exchange rate movements. These shocks might have led to a resurgence of inflationary expectations and greater upward pressure on prices. The avoidance of sharp depreciations of the exchange rate therefore received a good deal of weight in monetary policy decisions.

8. What monetary policy decision would avoid depreciations of the exchange rate?

"Secondly, if we can't borrow all of the funds we need from abroad to finance expenditures which we are going to make in Canada in any event, we should at least look at the advisability of the Bank of Canada itself financing some of those expenditures simply by creating the money."

That might be no more inflationary than borrowing the same amount of money abroad to finance the same expenditures, he said.

9. Comment on the view expressed in this clipping.

J. Forecasting the Exchange Rate

By narrowing the spread between short-term interest rates on commercial paper in Canada and the United States in line with the Canadian dollar forward rate, the bank stopped the arbitrage activity that was pushing up the Canadian dollar, Mr. Marshall said.

1. Explain how this arbitrage activity works.

Although they may think they have little in common, the conservative corporate treasurer and the brash, high-flying currency trader are alike in one important respect—they both speculate on the value of the C$.

2. How could a corporate treasurer be a speculator on the value of the Canadian dollar? How does he or she avoid undertaking such speculation?

Because it is impossible to predict accurately the timing of swings in the Canadian dollar, importers, exporters and others who have transactions in foreign currencies should hedge at least a portion of their obligations.	**3.** Explain in your own words what this statement means.
Floyd also analyzes another issue; the factors behind the recent decline in the value of the C$. That decline, which started about a year ago, saw the currency fall to US74.86¢ in July—a record low. Floyd says that a speculative blip can be ruled out as the reason because the currency was selling at a forward premium for most of the period	**4.** Explain the logic behind Floyd's ruling out a speculative blip.

K. The J Curve

A number of economists and business leaders say the increase in the value of the C$ has not yet had much impact on sales to the U.S. or on Canadian economic activity. But currency changes seldom have an immediate impact. Foreign buyers need time to react to currency-induced price changes.	**1.** What terminology do economists use for this phenomenon?
The problem is that it has taken longer than usual for the exchange rate to affect Canadian trade. Lawrence Schembri, an economics professor at Ottawa's Carleton University, noted that it generally takes six months to a year before the J-curve effect takes hold. This time, however, it has taken 1½ years.	**2.** What point on the J curve is being referred to by the phrase "takes hold"? Explain.

II-9 Stagflation

Economists used to think that there existed a trade-off between inflation and unemployment, represented graphically by the Phillips curve (shown by SRP_1 in Figure 11-9.1). This curve was thought to be downward-sloping because a rise in unemployment should dampen cost-push forces operating to sustain an inflation, and a fall in unemployment should strengthen these forces. The existence of this trade-off implied that the economy could "buy" a reduction in unemployment with an increase in inflation or "buy" a reduction in inflation with an increase in unemployment. All a policy-maker needed to do under these circumstances was to determine the character of his economy's Phillips curve, choose the point on that curve which was considered the least undesirable, and then adopt monetary or fiscal policy to move the economy to that chosen position. Throughout the 1960s this theory was regarded with some respect; policy-makers eventually subscribed to it and undertook policies accordingly.

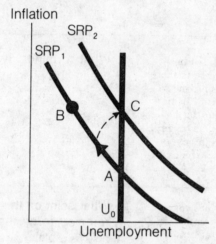

Figure 11-9.1

But these policies did not lead to the expected results; if anything, things seemed to become worse—the economy began to experience stagflation (high levels of both inflation and unemployment), a phenomenon the Phillips curve theory claimed did not exist. In their search for explanations of this phenomenon economists have considerably modified their conception of the Phillips curve. Regardless of whether or not a stable Phillips curve relationship existed in the late 1950s and early 1960s, everyone recognizes that the world we live in today is quite different, characterized by powerful labor unions, big business, and easily mobilized expectations. In recognition of this new world, the Phillips curve has been modified empirically by incorporating additional explanatory variables and theoretically by developing the concept of the long-run Phillips curve and the natural rate of unemployment.

The long-run Phillips curve is represented graphically in Figure II-9.1 as a vertical line over U_0, the natural rate of unemployment. This unemployment rate is the rate to which the economy gravitates in the long run. It is determined by people's tastes for leisure versus work, by the amount of frictional unemployment (people changing jobs or retraining) required as changing tastes and technology change the nature of what is produced, and by institutional phenomena such as minimum wage legislation and unemployment insurance programs. According to the theory of the long-run Phillips curve, any attempt to push unemployment below this long-run natural rate will be successful in the short run, but unsuccessful in the long run.

Graphically, if the economy is at point A in Figure II-9.1, where the short-run Phillips curve SRP_1 cuts the long-run Phillips curve, an attempt by the authorities to stimulate the economy will send the economy up towards B, along SRP_1, increasing inflation and decreasing unemployment. The real reason that producers increase output in response to this policy is that the policy caused prices to rise more than wages, making expansion by firms a profitable venture. After a while, however, labor will realize that prices have increased faster than they had anticipated they would, and will demand catch-up increases in their wages. Furthermore, they will now expect prices to rise at a faster rate and will build this expectation of higher inflation into their future contract demands.

This reaction by labor wipes out the profitability associated with producing at a higher level, so firms contract their output levels, returning unemployment to its original level. But because of the new, higher level of inflation expectations, inflation does not fall back to its original level. On the diagram, the higher expectations of inflation shift SRP_1 to SRP_2 (i.e., at each level of unemployment inflation is higher because of higher expectations of inflation—producers expect to be able to raise prices by more and so agree to the larger wage increases demanded by labor). The economy moves towards point C on the diagram, implying that in the long run the economy moves along the long-run Phillips curve rather than the traditional downward-sloping short-run Phillips curve. This means that there is no trade-off between inflation and unemployment, and that persistent government action to move the economy to a lower level of unemployment will only serve to push the economy to higher and higher levels of inflation, with no long-run improvement in unemployment.

All of this creates a policy dilemma. Aggregate demand policies (traditional monetary and fiscal policies) have no long-run impact on unemployment, suggesting that they should be directed to achieving a low level of inflation. This is done by moving the economy down the short-run Phillips curve, creating extra unemployment which eventually lowers inflation expectations. Unfortunately, this involves a prolonged period of high unemployment. Wage-price guidelines are often used to supplement this policy; they can help reduce inflation expectations more quickly, allowing the economy to avoid this period of high unemployment.

Most other policy suggestions in this context involve shifting the long-run Phillips curve (as well as the short-run Phillips curve) to the left by reducing the natural rate of unemployment. These policies are designed to eliminate or alleviate the imperfections in the commodity and labor markets. Examples of such policies are labor-retraining programs, programs to increase employment information or reduce the cost of obtaining such information, programs to increase labor mobility, the elimination of product and labor-market monopolies, elimination of legal minimum-wage rates, restructuring of unemployment insurance programs, and manipulation of government demand and taxation so as to alleviate demand in geographical or industry bottlenecks.

II-9A The Natural Rate of Unemployment

Low unemployment cause for concern: study

Dow Jones Service

OTTAWA

A new Bank of Canada study suggests that inflationary pressures are likely to develop on the Canadian labor market when the country's unemployment rate drops below 8 per cent.

The Canadian unemployment rate was 7.6 per cent in December. Last year it was under 8 per cent, except for January and August.

Some economists said the declining unemployment rate partly explains why the central bank has been pursuing a restrictive monetary policy for more than a year.

James Frank, chief economist at the Conference Board of Canada, said the Bank of Canada is using such indicators as the unemployment level in a "pre-emptive fight against inflation." He said the central bank wants to prevent another inflationary outburst such as occurred in the late 1970s and early 1980s.

The central bank's bank rate rose by more than a fifth of a percentage point Thursday, to 11.42 per cent, its highest level in three years.

Kevin Hayes, economist at the Canadian Labor Congress, said the unemployment rate shouldn't be used as an inflation barometer. Other countries have managed to keep both inflation and unemployment low, he said. He noted that Canada had low inflation in the late 1960s, when its unemployment rate was 3 to 4 per cent.

The Bank of Canada study, however, said the equilibrium rate of unemployment, or non-accelerating-inflation rate of unemployment, varies with changes in the labor force and labor laws. In the United States, most studies put the NAIRU rate between 5 and 6 per cent.

The study said Canada's comparatively generous unemployment insurance program has had a "significant" impact on the equilibrium rate of unemployment. It estimated that the rate was increased two percentage points by the introduction in 1971 of more generous unemployment insurance provisions. If assistance is available to unemployed workers, they are likely to wait longer for a better offer, raising the level of the NAIRU rate.

Robert Dussault, economist with the National Bank of Canada, also said Canada has special regional and structural labor problems that tend to push up its NAIRU rate. No matter how the economy performs, he said, unemployment will remain high in some regions of the country. Newfoundland's unemployment rate was 13.3 per cent in December, more than double the 5 per cent rate in Ontario.

The Bank of Canada report, written by David Rose, said measurement of the NAIRU doesn't lend itself to precision. He said there are demographic factors working to reduce the rate over the next few years.

(Reprinted by permission of Dow Jones News Service, © Dow Jones & Company, Inc. 1989. All Rights Reserved Worldwide.)

Note: Suggested answers to the questions below appear in the Appendix.

1. By means of a Phillips curve diagram, illustrate the statement made in the first paragraph.

2. Explain "why the central bank has been pursuing a restrictive monetary policy," as stated in the third paragraph.

3. Evaluate the statements of Hayes, in the sixth paragraph.

4. Give some examples of "changes in the labor force and labor laws" referred to in the seventh paragraph.

5. The acronym NAIRU appears three times in this article. What does it stand for? Explain clearly the rationale behind the name.

6. What do you think is the main reason why Canada's NAIRU is higher than that of the U.S.?

7. Give an example of a demographic factor which would reduce the NAIRU, as suggested in the last paragraph.

II-9B Unemployment versus Inflation

Time to focus on unemployment

JOBS, JOBS, jobs, candidate Brian Mulroney promised during the last election, and jobs, jobs, and more jobs have indeed poured forth: The ranks of those at work have swelled by more than one million since 1984.

From here on, however, things will not be so easy. It will not be enough to rely on economic growth to deliver jobs in bunches. This is not simply because the economy is expected to slow somewhat from the rapid advance of 1987. The evidence, rather, indicates that unemployment, at 8.1 per cent for the past two months, has settled in around a base level below which no sustainable rate of economic growth can take it. That is, the best estimates are that unemployment has reached its "natural rate."

This is an unfortunate and much misunderstood term. It does not mean in any way that this is acceptable, normal, or inevitable, let alone desirable. It means only that unemployment cannot be pushed below that level by expanding demand, which is to say by pumping up the money supply, without igniting inflation, and cannot be kept there without accelerating inflation even further. This is often said to be because higher wages "drive up" inflation. But both are functions of the same thing: growth in money supply beyond levels compatible with growth in potential output.

Far from fatalism or complacency, this is a call to action. Unemployment has remained above 7 per cent in Canada for the past dozen years, through tight money and loose, big deficits and small. It's a tragedy and a waste. Overall improvement in recent years, moreover, masks regional, occupational and generational disparities that tear at the social fabric.

It is the very diversity of the problem that suggests the roots of unemployment do not lie today in insufficient aggregate demand, a matter of pulling the right macroeconomic levers. Unemployment is chiefly a supply-side problem, requiring specific microeconomic reforms. It is the product of malfunctioning labor markets, and it is to the task of making them work better that the government must now turn.

---

1. What is the alternative name for the "natural rate" of unemployment described in the second paragraph?
2. Explain, by using a Phillips curve diagram, why "unemployment cannot be pushed below that level by expanding demand . . . without igniting inflation," as stated in the third paragraph.
3. Use the diagram from question 2 above to explain why it "cannot be kept there without accelerating inflation."
4. The last sentence in the third paragraph states a result which provides the rationale for what well-known policy?
5. What are the macroeconomic levers referred to in the last paragraph?
6. How would making malfunctioning labor markets "work better," as suggested in the last paragraph, be reflected on your Phillips curve diagram?

Short Clippings

Note: Suggested answers to the odd-numbered questions below appear in the Appendix.

A. The Natural Rate of Unemployment

Those figures coincide with the release of a Bank of Canada study that suggests that because of the structure of Canada's economy, inflationary pressures are likely to develop when Canada's unemployment rate drops below 8 per cent.

1. What name would economists give to this level of unemployment?

The damnable paradox is that we are also at "full employment," or at least at the "natural rate" of unemployment. This means only that we are at or near the lowest rate to which unemployment can be pushed by expanding demand without. . . .

2. Complete this clipping.

Unemployment rates in both Canada and the United States are sufficiently low that employers are having difficulty filling many jobs.
  The point at which red flags go up, economists say, is nine per cent in Canada and six per cent in the United States.

3. a) What name is given to the unemployment rates of the second paragraph?
   b) What danger do the red flags signal?

He favors a slow recovery, for example, partly because "the natural rate of unemployment" may turn out to be higher than anyone thinks.

4. a) What is the natural rate of unemployment?
   b) Explain why it is suggested that policy should push the economy out of recession slowly rather than quickly.

The traditional macroeconomic approach, flooding the whole economy with demand in hopes of floating any unemployed little boats, no longer applies. Unemployment today is more a matter of defective labor markets. So, make labor markets work better.

5. a) With whose name is the "traditional macroeconomic approach" usually associated?
   b) Why is this traditional approach no longer applicable? Use a diagram to illustrate your answer.
   c) How is the policy prescription of the last sentence portrayed on your diagram?

It is not enough simply to coast on growth, as the government has to date. Nor will NDP decrees of 1 per cent cuts in interest rates bring anything but more inflation. Nor yet do grand spending programs, whether the Grits' $5 billion roads-and-sewers plan or the Tories' multibillion megadoggles or whatever Ozymandian scheme the NDP devises, "create" jobs: they only redistribute them, from industry to industry, from region to region, from private sector to public sector.

6. a) Why will 1 per cent cuts in interest rates bring more inflation?
b) Why will the spending policies suggested merely redistribute rather than create jobs? What textbook macroeconomic theory is being employed here?

B. Interpreting the Phillips Curve

Some recent attempts to redo the Phillips Curve hypothesize that the relationship should be expressed in terms of unemployment and the *change* in the rate of inflation. One such effort along these lines in the U.S. concludes that, viewed in these terms, "there is no evidence the economy has changed in some fundamental way."

1. What rationale lies behind this new interpretation of the Phillips curve?

He postulates that "to reduce inflation by 1 per cent, we must endure unemployment 1 per cent in excess of "normal" [that is, in excess of 5 per cent] for one year."

2. Explain the reasoning behind this view.

Another school of thought suggests that the unemployment and inflation tradeoff is too simple-minded in any form. According to this view, productivity changes can explain to a large extent why there often appears to be no immediate relationship between unemployment and inflation.

3. Explain how productivity changes play this role.

What happened, instead, was that in the late 1960s the tradeoff curve shifted upward (more unemployment for a given level of inflation) rather than downward. Then in 1973-75, the old tradeoff was wiped completely off the map.

4. a) What tradeoff curve is being referred to here?
b) What would cause it to shift? Explain.

The response of the monetary authorities to any hint of inflation is to dampen demand by raising interest rates and slowing economic activity—that is, *by causing a recession*. It always seems to me that one of the great failures of the economic profession is that the only thing it seems to be able to recommend to eliminate inflation is a recession, but that's the way it is.

5. a) What diagram is used to illustrate this phenomenon?
   b) What role does the recession play in the economy's movement to a lower inflation rate?

But Michael Manford, vice-president and chief economist of Merrill Lynch Canada Inc., believes that the current high-capacity utilization and low unemployment rates mean trouble.

6. Where on a Phillips curve diagram would Manford place the current economy?

Anyway, less inflation means more unemployment. No longer, says Mr. Sprinkel. It used to, in the 1950s and 1960s. But the link disappeared in the stagflation of the 1970s. In the 1980s America has succeeded in lowering unemployment and inflation at the same time, and the council believes this can continue even though industry is now running at more than 80 per cent of capacity.

7. a) Explain the thinking of the 1950s and 1960s referred to in the clipping.
   b) Explain how "the link disappeared" as stated in the fourth sentence.
   c) Explain how inflation and unemployment could have been lowered at the same time, as stated in the last sentence.

A rapid recovery is needed because growth so far has been shakily based on a bit of returning optimism on the part of consumers who seem to have dipped into their savings to buy.

To be solid it needs growing real incomes—wages rising faster than the inflation rate—as well as enough business confidence to restart investment.

8. Comment on the extent to which the need for wages to rise faster than the inflation rate is exactly the opposite of what standard macroeconomic theory would claim is necessary for moving out of a recession.

Maxwell says the trouble is partly that governments have made mistakes but more because their policies don't match an economy in which everybody seeks protection from market forces.

And he suggests wages should be geared, not to prices but to real growth of the economy. Workers would then see their wages grow rapidly when productivity grew.

It wouldn't be easy to put wages on a new basis but it would be better, as Maxwell says, than the "present tragic output losses."

9. a) How are people seeking protection from market forces?
   b) In Maxwell's opinion, what is the main cause of the current recession?

C. Using the Phillips Curve

We're not likely to see a really tight monetary policy in Canada. Bank of Canada Governor Gerald Bouey made it clear in his annual report last week that the Bank is following an "intentionally moderate" monetary policy in order "to minimize the strains involved in adjusting to a less inflationary economy." He points to the "awkward economic fact that in the short run anti-inflationary policies tend to restrain output more than prices."

1. This is a good example of the way in which bureaucrats disguise harsh facts in flowery language. What is Bouey saying, in blunt terms?

"We cannot afford today to take any major risks with inflation. . . . If we let inflation get away on us again, even for just a while, the path back to price stability will be even more painful than it has been during the last few years."

2. Explain why the path is "painful."

"Walk, don't run." This is what Bank of Canada Governor Gerald Bouey was really telling Canadian businessmen and consumers last week when he announced that hefty increase in the bank rate to 9 per cent from $8\frac{1}{4}$ per cent.

Clearly, Ottawa's thinking is that what the economy needs now is not a speedy recovery from recession, but one that is slow, drawn out and, in many respects, painful.

3. What rationale would lie behind the last sentence?

Those policies were predicated on 1930s Keynesian assumptions that economic recoveries always run out of steam and at certain points need artificial stimulation of demand and fine-tuning to keep them running at acceptable levels. The evidence of the 1970s and beyond is that whenever governments stepped in to administer stimulative medicine, they triggered runaway inflation which finally had to be stopped with strong, painful doses of recession.

4. a) What name is usually given to the Keynesian concept of "artificial stimulation of demand"?
   b) Describe the macroeconomic theory used to explain the "evidence of the 1970s."

In his report to Finance Minister Marc Lalonde, he said that earlier successes of the use of financial stimulation "led to an excessive use of it and generated the Great Inflation of the 1970s."

5. Use a Phillips curve diagram to explain this clipping.

Bank of Canada Governor Gerald Bouey warned yesterday that economic recovery in the Western world could be short-lived if governments injected too much financial stimulus into their economies in an attempt to boost employment and output.

6. a) What is meant here by "financial stimulus"?
b) Explain why injection of too much financial stimulus would make economic recovery "short-lived."

But monetary policy cannot be tuned to real economic variables like growth or employment, not only because policy takes effect with long and uncertain lags, but because any commitment to real growth targets simply invites workers and business to increase wages and prices at will, in the knowledge that the central bank will "ratify" their demands via the money supply.

7. Use a Phillips curve diagram to illustrate a scenario reflecting the contents of this clipping.

"If stimulation is overdone . . . then I think the effect is likely to be perverse," Mr. Bouey told an Ottawa press conference at which his annual report was unveiled. "People will anticipate higher inflation because that is what they have seen in the past when stimulation has been overdone.

"They will use whatever opportunity they can to protect themselves by demanding higher wage increases, by pushing up prices, by requiring higher interest rates. Well, that's not going to help any recovery."

8. Use a Phillips curve diagram to describe the scenario structured in this clipping.

D. Policy for Stagflation

They seldom say, straight out, that the cure for inflation is to have enough people out of a job, enough factories idle, enough consumers unable to buy what is displayed in the stores.

They talk in a kind of code, about "restraint" and "gradualism" and "disinflationary demand policies."

1. Explain, with the help of a Phillips curve diagram, how this "cure" works.

Even if inflation was rising, could the central bank slow it without severely damaging the economy? The best answer is that monetary attempts to curb inflation have never been successful without causing a recession.

2. Use a Phillips curve diagram to illustrate the statement in the second sentence.

240 Part II *Macroeconomics*

Moreover, if a policy is to be credible enough for workers, consumers and investors to build their expectations upon it, then at the least it must appear feasible to them. Since the bank has not achieved zero inflation in the postwar era, why should individuals believe in such a policy now?

If a zero-inflation goal lacks credibility, then an attempt by the bank to achieve it will have the same sort of side effects as the policy to reduce inflation from 12 per cent to 4 per cent.

3. Use a Phillips curve diagram to illustrate the statement in the second paragraph.

"What it does mean is that over-all financial policy must be, and must be seen by a skeptical public to be, consistent with a continuing movement towards cost and price stability."

4. a) What kind of monetary policy would be "consistent with a continuing movement towards cost and price stability"?
   b) Why is it important that the "skeptical public" sees this consistency?

What is important for inflationary expectations is how the bank has responded to such shocks in the past. A central bank which has shown it will not accommodate inflationary shocks with monetary expansion will find it much easier to maintain a stable inflation rate—whatever its level—than one with a reputation for accommodation.

5. Explain the rationale behind this clipping. Use a Phillips curve diagram to illustrate your answer.

According to this theory, the U.S. recession-inflation debate boils down to whether the natural unemployment rate is higher or lower than the current 5.7 per cent level.

6. a) What is the theory being referred to?
   b) The recession-inflation debate is concerned with the question of what policy is appropriate. Explain how the answer to this question depends on "whether the natural rate is higher or lower than the current 5.7 per cent level."

"We are frequently pressed by people to do things that would involve giving up control over money creation without any apparent recognition on their part that that is what they are asking. We, of course, have to refuse."

7. Give two examples of such requests.

E. Wage and Price Controls and Guidelines

Let wage and price controls override the temporarily derailed market mechanism for 1-1½ years. This would give the adjustment mechanisms time to get back in working order and, more importantly, break the inflation psychology.

1. a) In what sense is the market mechanism "temporarily derailed"?
   b) What is the inflation psychology referred to?

"Instead of prices restricting consumption, it is shortages which operate." In other words, you cannot buy a thing, not because its price is too high but because it isn't there.

2. What macroeconomic policy does this clipping describe?

The second group contends that there wasn't a strong cause-and-effect relationship between money supply growth and inflation in the 1970s. The causes, they say, lie rather in the momentum brought about by a number of price shocks, notably grain and OPEC which produced inflationary expectations, particularly in the wage determination process. This momentum is hard to turn around once it gets going and the theory is that it takes government action-including an incomes policy or controls.

3. a) This group, advocating incomes policy or controls, seems to believe that it doesn't matter what is happening to the money supply. What do you think would happen if their policy were adopted but the money supply continued to grow at a high rate?
   b) What does your answer to a) suggest is the real difference between this group and the monetarists?

For example, prices and incomes policies have been tried on several occasions and in a number of countries, including Canada and the United States. The U.S. program, which was adopted in the second half of 1971, partly suspended the market system in that country, producing distortion and shortages while suppressing inflation instead of resolving it. Because of the failure of incomes policies wherever they have been tried, more prominence has recently been given to the possibility of broad-based indexing as a solution to the inflationary problem.

4. a) What is the difference between suppressing and resolving inflation?
   b) What is broad-based indexing? How could it be a solution to inflation?

Canada is one of the worst countries in which to try to make controls work because of its heavy foreign trade involvement, regional diversification and divided sovereignty.

Now is the worst time to apply controls because of the large and uneven changes in prices and wages that are occurring.

5. a) Why would heavy trade involvement make it difficult for wage and price controls to work?
   b) Explain the rationale behind the last sentence.

F. Supply-Side Economics

"It is not at all a choice between supply and demand," says economist Richard Dym. "For years, the scale has tipped toward controlling the economy by controlling federal spending and consumption. Now, it may begin to lean toward stimulating production through incentives."

1. What are the names usually given to the two types of policies towards which this clipping tips the scales?

Prof. Arthur Laffer's famous free-hand curve showing the effect of tax reduction on output, the magic logo of the supply-siders, was also not taken seriously by the profession. For some, Laffer was a figure of fun. Most others held that the Kleenex, paper napkin, or toilet paper on which, according to varying legend, the curve was first drawn could better have been put to its regular use.

2. Explain the Laffer curve. Why do most economists hold it in little esteem?

Inevitably, says Laffer, this economic growth will raise enough tax revenues to more than offset the original tax cut, thus shrinking the federal deficit and reducing inflation.

3. a) Explain how cutting taxes is supposed to stimulate economic growth.
   b) How can the "Laffer curve" be explained using Laffer's statement in this clipping?

Michael Spence of Harvard University is fearful of the high risks involved in some supply-side policies. "The Reagan people are currently arguing over what level of tax cuts should be implemented. What if they're wrong? What if the supply side doesn't respond enough to at least take care of the foregone revenues?"

4. Explain the meaning of the last sentence.

Supply-side economics, packaged and popularized recently as "Reaganomics," has become a big deal in the current policy debate over the fight against stagflation.

To counteract the phenomenon of persistent high inflation, unemployment and stagnant growth, supply-side policies emphasize increasing production to take the pressure off prices.

5. Use an aggregate supply-aggregate demand diagram to illustrate how supply-side policies could accomplish the results claimed in the second paragraph.

"If adverse supply shocks (escalating oil prices, crop failures) were so influential in creating stagflation in the 1970s, it is difficult to believe that we cannot create favorable supply shocks in the 1980s," argues economist Lorie Tarshis, in a paper prepared for a recent Ontario Economic Council conference.

6. How would these "adverse supply shocks" create stagflation?

Lipsey points to evidence that when demand is held too long below capacity output, "capacity itself may shrink (at least relative to the potential labor force) with disastrous consequences for the inflation-unemployment trade-off." This means that successive demand-induced recessions compound the problem and lead to ever increasing levels of inflation and unemployment.

7. a) Why would capacity shrink when demand is held too long below capacity output?
b) Why would this have disastrous consequences for the inflation-unemployment trade-off?

There is some suspicion that what cuts inflation in the short run may make the economy more inflationary in the long run.

8. Explain how this could be so.

On the surface, this sounds like a more rational and socially humane way of getting out of the current economic rut. Especially when the track record of fiscal and monetary restraint is so dismal, leading many to argue that it is inappropriate for dealing with stagflation in a democracy. (Watching Britain indulge in its severe monetary restraint experiment may settle this once and for all.)

9. What is the "severe monetary restraint experiment"?

II-10 Government Deficits

There are three basic ways of financing an increase in government spending: collecting taxes, selling bonds, and printing money. If the government raises taxes to finance its extra spending, the higher taxes lower disposable incomes, decrease consumption, and thus largely offset the increase in aggregate demand resulting from the higher government spending. If the government instead sells bonds (this is the financing method usually assumed, when talking of an increase in government spending), the money used to buy these bonds cannot be used to finance other spending, so other spending decreases, again partially offsetting the increase in demand due to the higher government spending. (This can also be explained by appealing to the higher interest rate created by the government bond sales, causing interest-sensitive demand to decline.) When an increase in government spending is financed by the third means (printing money), however, the stimulating effect of higher government spending is aided by the concomitant increase in the money supply.

The first two means of financing an increase in government spending described above involve the "crowding-out" phenomenon: The financing means crowds out other spending, resulting in a smaller policy impact. The monetarists (those feeling that monetary policy is stronger and more reliable than fiscal policy) make much of these crowding-out effects, and claim that the impact of fiscal policy when financed by printing money is due to the increase in the money supply rather than to the increase in government spending.

Whenever the government brings down its budget it announces what its anticipated deficit, or cash requirement, will be. (Of course it could expect a surplus, but in modern times this doesn't seem to happen. The discussion to follow is in terms of a deficit.) The anticipated deficit is the difference between planned government spending and projected tax revenues, a difference that must be financed by the two remaining financing means. A great failure of modern budget presentations is that they do not reveal how much of their deficit is to be financed by selling bonds and how much by printing money. As a result, the degree to which the budget can be considered expansionary (or contractionary, if there is a surplus) cannot be gauged, since the impact on the economy of any change in government spending depends as much on the means used to finance that change as it does on the size of the change.

To the uninitiated, this neglect is compounded by the fact that the government does sell bonds to cover all of its projected deficit. Invariably, however, a large quantity of those bonds are bought by the Bank of Canada; this is what is meant by the expression "printing money." If you and I buy a government bond, our bank balance falls and the government's bank account rises—there is no net change in total deposits in chartered banks and thus their reserves remain unchanged. But if the central bank buys a government bond, the government's deposits in chartered banks increase with no offsetting decrease—the money supply increases.

In the early and mid-1980s U.S. and Canadian government deficits climbed to extremely high levels. The articles in this section describe the circumstances surrounding these deficits, review the concerns they created and assess related policy suggestions. All three are interesting and informative. The third article stresses that policy suggestions with regard to these deficits must assess both their costs and their benefits, a useful general perspective for this problem.

II-10A Debt Problem

Total government debt is real problem

Bonds

BY JOHN GRUNDY

The damaging effect to the marketplace created by the persistence of heavy government borrowings is an ever-present concern for bond investors. Although provincial governments are steadily increasing their portion of total public debt, it is Ottawa that receives most of the verbal abuse. As the largest new-issue borrower in the market, the federal government's heavy demands are seen as a major impediment to sustainably lower interest rates.

A fear is that the solution to Ottawa's seemingly insatiable appetite for funds will be debt monetization. As a result, most observers expect inflation-adjusted interest rates to remain historically high.

Debt monetization occurs when the Bank of Canada credits federal government accounts with new funds, essentially with the stroke of a pen. The most direct method is for the federal government to issue bonds and sell them solely to the Bank of Canada. Such an exercise, if not offset with other actions by the central bank, dilutes the value of the dollar and so pushes inflation higher.

The Bank of Canada has shown not the slightest tendency to accommodate the monetization "solution." In fact, the bank has become recognized as a leader in adopting prudent monetary practices that have not encouraged any additional fiscal folly to that already existing. Nonetheless, the fear of future monetization remains.

Encouraging the concern is the persistence of huge government budget deficits. If left unchecked, they will become so overpowering that the ability of the private sector to acquire reasonably priced capital will be severely curtailed.

---

1. Explain why "the federal government's heavy demands are seen as a major impediment to sustainably lower interest rates," as stated in the first paragraph.

2. What is debt monetization? How is it a "solution to Ottawa's seemingly insatiable appetite for funds"? Why should we "fear" it? (second paragraph)

3. What are "inflation-adjusted interest rates"? (second paragraph) By what name do economists refer to them?

4. Explain the meaning of the first sentence of the fourth paragraph.

5. Explain how "the ability of the private sector to acquire reasonably priced capital will be severely curtailed" as claimed in the last paragraph.

6. What is the "damaging effect," referred to in the first sentence of the article, that is an "ever-present concern for bond investors"?

Putting a ceiling on deficit right policy for these times

William G. Watson

MONTREAL

LAST WEEK IN *The Post* I argued that the employment-creating effects of increases in the federal deficit are probably overrated. The reason is that a bigger deficit very likely "crowds out" other forms of demand.

The government's entry into the capital markets as a major borrower puts upward pressure on interest rates. This either discourages private investment directly or causes a reduction in exports as buoyant interest rates pump up the value of the C$. Another possibility is that consumers will react to the higher future taxes the deficit requires by saving more now to prop up their future after-tax incomes, and these higher savings depress current demand.

In assessing the *costs* of the deficit, economists naturally ask whether increasing the deficit reduces anyone's income. More often than not, the answer is that higher deficits probably do depress future generations' income. In the case of crowding-out of private investment, for instance, future income is lower because the capital stock successor generations inherit from us is smaller. How big is the loss in future income? It's equal to the real rate of return on whatever private investment is squeezed out by the government's entry into the capital markets.

As suggested last week, however, small, open economies don't usually have to worry about physical crowding-out of this sort, since they can get all the capital they need from the international capital market. Thus neither the future capital stock nor future Canadian output is reduced. But this doesn't mean the deficit is costless: the return on whatever investment is financed out of foreign savings has to be paid to its foreign owners. This reduces the *income* future Canadians receive even though it doesn't reduce the *output* they produce. On the other hand, if both employment and the level of workers' wages are important concerns, foreign investment will be good for both.

In the admittedly extreme case in which people respond to an increase in the deficit by saving more, then there is no loss of future income at all, since private investment is not crowded out and the future capital stock is exactly what it would have been without the deficit. Whatever loss the deficit causes occurs right now, if government spending buys things that provide less satisfaction than private consumption goods would have. But if so, the real beef about the deficit is simply that it enables the public sector to become too large. From this point of view, whether extra revenues are raised by taxation or borrowing is entirely immaterial.

Future returns

Of course, if the government uses borrowed funds to invest in assets—hospitals, schools, roads, dams—which themselves have large future returns, then any costs of crowding-out will be at least partly offset by the boost such investments give to future income. This raises the often-overlooked point that the government shouldn't necessarily try to balance its budget over the business cycle, as is often recommended. If "profitable" public sector investments are available, then they should be undertaken whatever the current rate of unemployment. The same reasoning suggests that what we really want to be looking at is not just the government's debt, but its net worth.

None of the costs I've described sound very impressive. So—it might reasonably be asked—why not go ahead and expand the deficit? Even if the potential reduction in unemployment isn't as great as many people argue, at least it's movement in the right direction. To state the problem differently, if the costs of the deficit really aren't all that great, why is almost the entire business community volunteering—quite uncharacteristically—to pay their taxes now rather than later (which is what any early reduction of the deficit would require)?

What most worries potential investors about the deficit is the (albeit remote) possibility that things will come unhinged. As the ratio of debt to GNP rises, so, too, does the expectation that government ultimately will retire the debt by printing money. And if we do have another Great Inflation, both the investment it discourages and the policy-induced, inflation-fighting Great Recession that inevitably will follow it are likely to do considerable social damage.

Even if such worries are in some sense irrational, policymakers cannot afford to ignore them: irrational or not, they have real effects on people's behavior. Thus the bet is that at this stage private investors will be less discouraged by higher taxes and/or lower public expenditures than they would be by continued growth of the debt.

The risk here is that fear and loathing in fact would not set in until the debt rose to 80 per cent or 90 per cent of GNP—it's currently about 45 per cent—so that forgoing potentially fruitful stimulation is just excessive, wasteful caution. But that's a little like betting that if we dismantled our nuclear weapons, there's a good chance the Russians would do the same: we may be right, but the downside loss from being wrong is very large.

The federal deficit currently stands at about 8 per cent of GNP. With the unemployment rate at about 11 per cent, acting quickly to eliminate the deficit probably isn't wise. On the other hand, making clear that 8 per cent is about as high as you are willing to go makes considerable sense. Finance Minister Wilson's mini-budget was designed to convey this message. As such, it is probably the right policy for these times.

Note: Suggested answers to the questions below appear in the Appendix.

1. Explain the logic behind the last sentence of the third paragraph.
2. Explain the logic behind the last sentence of the fifth paragraph.
3. What will happen if the government retires the debt by printing money? Who will end up "paying" for the deficits if this route is followed?
4. If real interest rates are determined internationally, explain how a Canadian deficit causes "crowding out."

Short Clippings

Note: Suggested answers to the odd-numbered questions below appear in the Appendix.

A. Monetizing the Debt

A significant difference will exist between the consequences of federal as opposed to provincial borrowing if the federal government borrows by selling bonds to the Bank of Canada. It alone can do so—the Bank buys only Government of Canada bonds.	1. What is this significant difference?
First: If government debt is growing faster than the economy, real interest rates must rise, over time, unless private-sector debt growth provides a corresponding offset. Real interest rates must rise to induce the public to hold the additional government debt. If real interest rates do not rise, the public will, presumably, continue to spend the same percentage of its income as before, and there will not be sufficient new savings to fund the additions to the government debt.	2. What would happen if the government tried to avoid the rise in the real interest rate by selling the debt to the Bank of Canada instead of to the public?
Second: It is the rare consumer or investor who thinks about it who would not conclude that where government debt was growing faster than GNP, tax increases and/or accelerated inflation could not be far off.	3. Why is accelerated inflation an alternative to taxation?
There is no direct connection between the deficit and inflation. If there were, why did inflation hit 14.2 per cent in 1948, a year the government had a spending surplus? And why has inflation continued to fall the past two years while the deficit has been rising to about $35 billion this year?	4. a) What is the logic behind the view that there is a connection between the deficit and inflation? b) Do the facts cited in this clipping contradict this view?
But, it might be speculated, there is a more important reason for the central bank action on the bank rate. The increase seems to be a message to the federal government that the Bank of Canada is not satisfied that the mix of fiscal and monetary policies is sufficiently stringent to bring inflation under control.	5. The clipping suggests that the hike in the bank rate is a message to the federal government to decrease its deficit spending. What power has the Bank of Canada to back up this message should the government not decrease its deficit? What would happen if it did not decrease its deficit? (Hint: How does the government finance its deficit?)

B. Deficits and Interest Rates

Another criticism that Walker levels at deficit budgets is that they hamper the private sector's ability to raise capital.	1. a) How do budget deficits hamper the private sector's ability to raise capital? b) By what mechanism does this occur? c) How can the Bank of Canada circumvent this? At what cost? Explain.
One is whether the government can steel itself to bring its fiscal policy into line with the Bank of Canada's monetary policy. Will the government be able to borrow the funds it will need to cover this year's deficit—$11.6 billion or more—out of the existing money supply, which Mr. Bouey is trying to restrict?	2. How could the government ensure that it will be able to borrow the needed funds? Why may it not be willing to do this?
One valid reason for raising the bank rate was the prospect of a refunding issue: A $436-million issue matures on April 1, and the Government is expected to seek some new money, in excess of its refinancing needs.	3. If the government must refinance $436 million, why doesn't it get the Bank of Canada to wait until after the refinancing before increasing interest rates, so as to minimize its interest costs?
U.S. Federal Reserve Board Chairman Paul Volcker told senators Thursday the huge federal budget deficit is causing "disturbing pressures" on interest rates, developing countries and money exchange rates.	4. Explain how the huge federal deficit is causing these disturbing pressures. To what extent could huge federal deficits in Canada be said to cause similar pressures?
High interest rates, not the deficit, are the country's major problem. It is high interest rates that have caused unemployment which in turn has cut into government revenues and forced up spending, adding to the deficit.	5. How would you critique this statement?
Prices in all areas of the Canadian bond market traded in a narrow range for most of the week, but jumped sharply yesterday morning in reaction to the news that the U.S. Senate voted to approve specific deficit reduction measures.	6. Explain the rationale behind this report.

C. Crowding Out

About the only benefit ever claimed for the deficit is that it will bring an increase in the level of aggregate demand and therefore a reduction in unemployment. In recent years, however, this very Keynesian notion has begun to be doubted by many economists.

The main worry is "crowding out."

1. Explain in your own words how crowding out comes about.

Economists at the seminar agreed that the U.S. deficit of about US$180 billion is severely straining domestic capital markets because private-sector capital demand is proving to be exceptionally buoyant. The result is a classic "crowding-out" situation.

In Canada (where the federal government's financial requirements will be an estimated $26 billion for fiscal 1984-85), crowding out is currently less of a problem, said Michael Manford, vice-president and chief economist at Merrill Lynch Canada Inc. The reason is that private-sector demand for capital is rather weak and the Canadian personal savings rate is far higher than in the U.S., which means that Canadian borrowers have a relatively larger pool of funds to draw on.

2. a) What is meant by "a classic 'crowding out' situation"?
b) How would you qualify Manford's statement in recognition of the fact that Canada has a small open economy?

Of course, if Ottawa needed less cash it would undoubtedly ease the strain on the provinces and the corporate sector.

3. Interpret this statement. What is it that determines whether Ottawa or others get the cash?

He continued: "Because the federal deficit is largely the result of transfer payments over which the Treasury Board has only modest control, the governments of Canada are forced into the debt markets on an ever-increasing scale, where they soak up available funds, making it even more difficult for the private sector to find the finance it needs."

4. How can the private sector circumvent this competition from the government? What impact on the economy would result from their doing this?

Also, the financing room freed by reduction in federal deficits is not necessarily used to expand productive capacity. As inflation persists, borrowing to buy existing "inflation hedge" assets crowds out other financing needs in a way that is not totally different from federal borrowing.

5. Explain this clipping in your own words.

There is no pleasant federal budget package that can bring down interest rates. A package is only meaningful if it carries the threat of recession, in the same way as monetary policy is meaningful only if it carries the threat of recession.

I refuse to believe a change in the deficit will produce a parallel change in total borrowing. One should expect that as the deficit declines, private borrowing will rise—that they operate as mirror images. The budget deficit is an organic part of the whole.

6. a) What does this person believe is the magnitude of the interest elasticity of the demand for investment funds by the private sector?
b) According to this logic, why would creation of a recession lower interest rates?

Q: So you reject the view put forward by Paul Volcker, chairman of the Federal Reserve Board, that a deficit reduction of US$50 billion would lower short-term interest rates by one percentage point?
A: I think he is sorry he said that. What I believe he meant to say is this: If the government's borrowing is US$50 billion less and private borrowings don't increase, then interest rates would be lower. In that sense, he is correct.

7. a) Explain the rationale behind Volker's view.
b) Evaluate the answer, given the question.

D. Deficits and the International Dimension

But while open economies don't have to worry about crowding out of investment, they do have to worry about crowding out of exports. The links between government deficits and exports run as follows. . . .

1. Provide the explanation of the link between deficits and exports.

Crowding out is based on the idea that there is just so much money available for loans. In other words, there is a single pool of cash from which money is drawn to cover government deficits, to expand business, to build houses and to buy such things as cars and furniture.

2. What is misleading, in the case of Canada, about this explanation of crowding out? How would you alter the explanation?

Lalonde wants to raise the stakes by increasing the already massive federal deficit.

He feels it can grow larger than the current estimate of $23.6 billion without inflicting further damage on the economy or pushing up interest rates.

3. How would you defend the view that increasing the deficit will not raise interest rates?

"If the minister had showed his measures could get the deficit next year below $30 billion, instead of $32 billion, I think that would have sent a signal to the financial community and we could have looked forward to more substantial interest rate reductions, like an extra one percentage point," says C.D. Howe policy analyst Ted Carmichael. "I believe the stimulative effect of this would have outweighed the costs of all cuts, because deficit reduction and employment go hand in hand."

4. a) How would you argue against the statement that a reduction in the deficit could lead to a substantial interest rate reduction?
   b) In what sense do deficit reduction and employment go hand-in-hand? In what sense is the opposite the case?

The U.S. can expect to get lower interest rates through a reduction in its budget deficit whatever happens to other countries' interest rates. That is not true to the same extent for other countries because even if they curb their deficits they will still be subject to the pervasive effect of U.S. interest rates.

5. Why is the U.S. a special case in this respect? Explain the mechanism through which other countries are prevented from lowering interest rates by curbing their deficits.

First, a rise in the ratio means that the current generation is consuming goods and services at the expense of future generations. In a small, open economy such as Canada's, increased federal borrowing results in a capital inflow and allows us to consume more than the Canadian economy produces.

6. The ratio being referred to here is the national debt—national income ratio.
   a) Explain how we can "consume more than the Canadian economy produces."
   b) Explain why this is done "at the expense of future generations."
   c) This clipping is assuming that Canada is borrowing from foreigners to finance consumption expenditures. Explain how borrowing for some other purpose might invalidate the argument that this is done "at the expense of future generations."

E. Structural Deficit

The arithmetic is straightforward. If growth falls one percentage point below the government's 3 per cent forecast in 1989, the deficit would widen by about $1.5 billion.

1. How can growth affect the government's budget deficit?

For 1984, for example, the structural deficit is about $11 billion (out of Marc Lalonde's $30 billion deficit) if the normal growth of the economy is put at three per cent. But it's calculated at about $14 billion if the economy is only capable of a normal growth of 2.6 per cent.

So a huge difference of $3 billion in the structural deficit results from a change of just four-tenths of one per cent in what is considered "normal" growth for the Canadian economy.

2. Explain the connection between growth and the structural deficit.

The fact that both deficits and unemployment have increased together since 1979 proves that fiscal stimulus doesn't work.

It's true that Canada's jobless rate rose to 11.3 per cent in 1984 from 7.4 per cent in 1979, while the federal deficit increased to 7 per cent of GNP from 3.5 per cent. But the inference about the ineffectiveness of fiscal stimulus is nevertheless wrong, because. . . .

3. Complete the argument about to be developed here.

Both the council and the private economic research group, the Conference Board of Canada, said last week that the deficit has bottomed out and will begin to rise next year.

The council predicted economic growth would slow this year to an after-inflation rate of 2.5 per cent from a robust 3.7 per cent last year. And it forecast that growth would slow again in 1989 to 1.8 per cent and in 1990 to 1.1 per cent.

4. What is the connection between the first and second paragraphs of this clipping?

Full employment deficits: on a cyclically adjusted basis the federal budget deficit in 1980 was $6.5 billion, or 2.1 per cent of GNP. That is to say, had the economy been operating at "average" capacity in 1980 there would still have been a deficit of major proportions.

It was inconceivable in Keynesian theory that governments could have major deficits at full employment.

5. a) Explain exactly how you would adjust the regular deficit to get this full employment deficit. Which deficit would be larger?

b) What is it about Keynesian theory that gives rise to the statement in the second paragraph?

The structural deficit—that part of the deficit which would exist if the economy was at full employment—is less than one-third of the total deficit. While this may be justified currently to support economic recovery, there are longer term consequences—particularly if the ratio of public debt to GNP rises through time. Accordingly, long-term fiscal planning must aim to stabilize the ratio of debt to GNP at an acceptable level.

6. Does this argument suggest that the structural deficit should be zero? Explain why or why not.

F. Keynesian Policy

The Conservative government's deficit-reduction policies are "eerily reminiscent of the 1930s" and will only drive the economy deeper into recession and push up unemployment, says a report issued Monday by a private think tank.

1. How could deficit-reduction policies "drive the economy deeper into recession and push up unemployment"?

However, there are some important differences in the relative situations in the two countries that suggest that the urgency of reducing the federal deficit right now is not as great in Canada as it is in the U.S. This conclusion follows from Canada currently being much further below its potential output than is the U.S.

2. Explain the rationale behind this view.

Though most of this money came from borrowing rather than government revenue, each case represents a political decision to spend money on job creation and economic development. Such projects admirably suit the government's purposes, by providing the basis for riding out the economic crisis as well as building the infrastructure needed to support future economic growth.

3. How does this clipping reflect a Keynesian view?

As to how the federal deficit would be cut back, the economists rule out tax increases. Instead, the cuts must come on the expenditure side with the average growth of federal expenditures—including interest payments on the national debt—kept below average growth of GNP.

4. Why would these economists choose to reduce the deficit by spending cuts rather than by tax increases? Hint: The answer reflects their subjective values.

On a more fundamental level, a change in the behavior of elected officials is required before we can achieve more stable debt levels. Politicians must begin to acknowledge in public that a decision to spend is a decision to tax.

5. Explain in what sense a decision to spend is a decision to tax, and in what sense it is not.

II-11 Free Trade

The benefit of free trade is that in the long run it will increase productivity and thus raise the standard of living in Canada. Less productive businesses, currently protected by tariffs, will disappear and be replaced by growth in industries in which we have a comparative advantage. Since there is no conclusive evidence that free trade will affect the natural rate of unemployment, it is not possible to predict what impact, if any, free trade will have in the long run on unemployment. Adjustment of the Canada-U.S. exchange rate will in the long run eliminate any short-run tendency for one country to gain more than the other.

There are two main economic costs of free trade. First, in the short run certain sectors of the economy will be forced to adjust. Many workers will have to retrain and find new jobs, meaning a temporary loss of income. Second, although in the long run Canadians as a whole will be better off, some Canadians will be worse off. Those owning industries protected by tariffs, and those earning wages that are exceptionally high due to tariff protection may suffer a decrease in their standard of living.

It is understandable that those people in our economy who are forced to make short-run adjustments and those forced to experience a long-run fall in their standard of living would object strongly to free trade, just as the proverbial buggy-whip manufacturers objected to the advent of the automobile. It is also understandable that they would fortify their objections by appealing to non-economic arguments such as the possibility of Canada losing her cultural sovereignty. How relevant are these non-economic arguments? How much sympathy do these people deserve? Should they be compensated somehow?

These are the questions that made the free trade debate such a prominent issue in the 1988 election. Unfortunately, the free trade debate and its newspaper coverage tended to focus on emotive issues, mainly political, with the economic issues playing only a background role. The pro and con articles that we have placed in this section reflect this. There are no specific questions for the two longer articles; rather, they are included to serve as input to classroom discussions of the free trade issue.

Free Trade: Pro

What kind of country is this anyway?

By Peter Cook

Let us step outside the arena in which the Mulroney-Turner cockfight is being held. Everyone inside has wagered something on the outcome. So they are all stomping and screaming for their champion to win.

Whenever an objective measure of what people think about Canada's election is brought forward—such as the serious disruptions rocking the financial markets and the dollar—it can be discounted.

The messenger is biased because he is a stooge of one camp or the other. Therefore, the message is discredited.

On such a basis, on free trade, everyone and almost every rational argument can be pulled to pieces. Liberal Leader John Turner—who has most to oppose—has gone furthest along this road. But he is helped by the suspension of belief that greets much of what Brian Mulroney and Ed Broadbent say. Business people and labor unions and trade negotiators and pressure groupies and social activists and Bay Street and Wall Street and, most especially, Ronald Reagan when he talks of the free-trade deal, are all without exception guilty of grinding on axes—or, more importantly, can be said to be.

This being so, it is worth looking at what knowledgeable foreign commentators, who have little interest in the outcome, are saying about Canada's vortical election. Okay, they too can be wrong or biased. But there is no doubt their opinion will influence what happens after Nov. 21.

First, the Lex column in the Financial Times of London on the subject of how international investors view Canada:

"After several years of underperformance, the world's fourth-largest stock market (Canada's) seemed poised to break out—until a clutch of opinion polls suggested that the ruling Conservative Party would either not return to power, or would do so without a large enough majority to force through the free-trade pact with the United States.

"To judge by the sharp drop in the Toronto stock market and the Canadian dollar, this came as a considerable shock to international investors. Three weeks ago, Canada could boast of rapid economic growth, low inflation, and a willingness to open up its markets. Now it is perceived as being in danger of retreating into its shell.

"If it passes up this opportunity to abolish tariffs with its biggest trading partner, it may not get another. And, if the world continues to move toward trading blocs, Canada could be left out in the cold."

Second, an editorial from The New York Times on the free-trade agreement:

"The overriding mutual benefits of unfettered commerce are competition, lower prices, more growth and more jobs. Canada's growth in the past four years has been impressive, but a North American trade partnership would help both partners face the rising challenges of Japan and Europe. Even so, there remains in Canada a powerful point of pride. Canadians have always worried that closer economic ties would overwhelm Canada's identity. They don't want to be Americanized.

"Opponents of the trade agreement play to these sentiments with outlandish fears—that Canada's generous health and social benefits would be attacked as 'unfair' subsidies to Canadian workers and that America could drain Canada's water supplies. The agreement poses no such threats. Its likely effect will be greater prosperity for both countries, thus bolstering Canadians' pride in Canada in tangible ways.

"At a time of rising protectionism, especially in the United States, achieving a free-trade agreement of such dimensions was nothing short of remarkable. If Canadians now choose to reject this opportunity for guaranteed free entry to the world's largest market, that's their business. And their loss."

Third, an excerpt from an article published in The Economist as long ago as June—just ahead of the Toronto summit—which has nothing, and everything, to do with this election. It was co-written by 10 distinguished financial and political figures including Paul Volcker, Etienne Davignon, Henry Kissinger, Lord Roll and (for Canada) Gerald Bouey. It states in one passage:

"There is no such thing as economic independence in today's world. Losing one's economic independence is not to be feared. It has already happened and has in fact produced much economic progress.

"Our countries (the Group of Seven, including Canada) have only two choices: accept the reality of today's and tomorrow's world through co-operative macroeconomic policies to correct imbalances and increase world economic growth and welfare—or accept that the price of fruitless efforts to follow more nationalistic policies will be the reduction of our own and others' standards of living. Commercial alliances, co-operation between enterprises and increased direct investment would give a much more concrete meaning to the concept of interdependence."

This is advice and commentary that can be accepted or rejected. But Canadians should be aware that, on free trade, international perception of what the country does far outweighs the limited economic effect of the trade agreement itself. Rightly or wrongly, this election is seen as a test of whether we want to join the world or turn our back.

The choice would matter less if Canada—with its small, open economy and huge public debt—was not desperately dependent on the world's good opinion.

Free trade could cost us 'Canadian way' of doing things

By Jack Gibbons

THE CANADA-U.S. free trade proposal has a bluechip cast of supporters. As a result, many commentators have concluded that the critics must be a motley crew of vested interests, economic illiterates, narrow nationalists and reactionaries. Is it really this simple? Is it impossible to prepare a rational brief against bilateral free trade?

VIEWPOINT

According to Richard Lipsey, one of our preeminent economists, the best empirical studies suggest bilateral free trade would eventually raise our national income by 1.3-7 per cent.

However, the 7 per cent upper bound estimate is unreasonable—it's based on out-of-date tariff levels and highly optimistic estimates on the productivity benefits flowing from industrial rationalization. Furthermore, to put these numbers in perspective, it is worth noting that since the Second World War our annual rate of economic growth has average 4.1 per cent. Thus, at best, free trade will give us the equivalent of approximately one year's rate of growth.

In addition to tariff elimination, it is asserted that a trade deal will give Canadians guaranteed access to the U.S. market. That is, U.S. trade laws (anti-dumping, countervail) could no longer be used to restrict Canadian exports (e.g., softwood lumber). Unfortunately, the U.S.-Israel Free Trade Agreement of 1985 demonstrates the Americans will not make such a major concession.

Even though Israel accounts for only 0.5% of U.S. imports it is still subject to U.S. anti-dumping and countervail legislation. How can anyone seriously suggest they will be more generous to the country that is responsible for 15 per cent of their imports?

The cost of a bilateral free trade deal is a loss of sovereignty. To the free trader, this is not a serious objection. As long as we are richer, albeit slightly, who cares about national sovereignty? The response of the majority of Canadians for more than 100 years has been quite simply that material riches are not everything. Canadians have wished to remain masters in their own house so that they can create a unique society in the northern half of North America.

To the U.S. government, free trade means all economic decisions are made by the marketplace. The role of government is merely to ensure there is a level playing field on which the contestants can compete eyeball to eyeball. This is not the Canadian way. We believe it is legitimate for governments to intervene in the marketplace to promote social justice, to enhance our cultural sovereignty, to protect our environment and to stimulate employment. For example, the federal government provides regional development grants to create jobs in less advantaged areas: Newfoundland fishermen receive UIC benefits during the off-season. Our provincial governments have created marketing boards to insulate farmers from the roller coaster vagaries of unregulated commodity markets. The federal government has legislated Canadian content requirements for TV and radio and placed restrictions on the foreign ownership of Canadian corporations. To Americans, these programs and policies are unfair export subsidies or nontariff barriers. They will demand their elimination as a prerequisite for a bilateral free trade deal.

Canada-U.S. free trade will also severely circumscribe the federal government's economic policy options. One of the most effective ways to increase employment is to lower the interest and exchange rates by increasing the money supply. A lower interest rate creates jobs by stimulating business investment, residential construction and the purchase of consumer durables. A fall in the value of the C$ increases employment by making our exports more competitive and by discouraging imports.

No doubt somewhere down the line the Americans will conclude medicare is an unfair export subsidy or an untoward interference in the market. Yet, conceding to all these demands will probably not be sufficient to satisfy the Americans. No less an authority than Simon Reisman, Canada's chief trade negotiator, has stated that to clinch a satisfactory deal it might be necessary to divert water from James Bay to the American Midwest. The environmental disruption that would flow from a $100-billion water export project is incalculable.

Finally, by increasing our economic dependence on the U.S., free trade would make the Canadian government even more reluctant to pursue an independent foreign policy.

If Canadians still believe Canada is a unique country worth keeping, they must once again say no to Canada-U.S. free trade.

JACK GIBBONS *is a Toronto economist*

Short Clippings

Note: More free trade short clippings appear in Section I-9. Suggested answers to the odd-numbered questions below appear in the Appendix.

Over the past 18 months, the C$ has gained nearly 10 per cent vs the U.S. greenback and still appears to be rising. Jim Frank, Conference Board of Canada chief economist, says this is like being hit by all the disadvantages of free trade and none of the advantages.	1. Explain why this is "like being hit by all the disadvantages of free trade and none of the advantages"?
"If the free-trade agreement is finally ratified by the Canadians, it will be important to eliminate the exchange rate advantage gained by Canadian producers over their U.S. counterparts in the period 1976-86," Mr. Trowbridge wrote. "There is a belief in the United States that the Canadian dollar should be at par and that anything less is a subsidy," said Douglas Peters, vice-president and chief economist of the Toronto-Dominion Bank.	2. Explain why this American belief is unfounded.
Canada would be an attractive place to invest without corporate taxes and with a fluctuating dollar to keep wage costs competitive relative to the U.S.	3. How would a fluctuating dollar keep Canada's wage costs competitive with those in the U.S.?
Tell them 200,000 textile workers will be unemployed because of unrestricted Korean imports and you'll hear a few tut-tuts. Not so, they say. Koreans can't export clothes to Canada without imports of Canadian goods to pay for them so free trade will mean a readjustment of jobs not a loss of them.	4. Suppose the Koreans sell the Canadian dollars earned through exports for U.S. dollars and import from the U.S. instead of Canada. Is the argument presented in this item still valid? Explain why or why not. (Assume a flexible exchange rate.)
Governor Bouey goes on to set up a challenge which the government ought to confront. To quote him: "It always seems to me that anyone who advocates a deliberate depreciation [of the dollar] ought to start there. He ought to say, 'what I really want is a reduction in real wages, a depression of real income.' You don't hear that very often."	5. Explain the logic behind this reasoning. How would an advocate of deliberate depreciation respond to this?

While it may help exports temporarily, the fact is that the cheaper C$ impoverishes Canada and Canadians much in the same way that inflation impoverishes people within the boundaries of a country.

If we can sell more goods and services abroad, how can that be?

6. How would you answer this question?

Appendix: Suggested answers to selected questions

I-1A Oil

1. The demand for natural gas shifts right (left) as the price of oil rises (falls). This is the key to answering this question.

2. They will have to abide by their agreement to artificially restrict the supply, shifting the supply curve to the left.

3. The excess demand created by the rightward shift of the demand curve over time due to population and income growth.

4. In excess of the rate of inflation.

5. The price at which unconstrained supply and demand match.

6. One in which there is a tendency for the price to fall.

I-1B Newsprint

1. The demand curve for newsprint has shifted to the right due to the recovery in the U.S. and Canada.

2. The demand curve shifts up by 5 per cent since the demand at a given price is in fact the demand associated with a price 5 per cent lower.

3. In the third and fourth-last paragraphs it is clear that the industry is meeting the higher demand by operating at higher capacity rates. Thus it is moving along its supply curve.

4. His definition seems to require that the industry be operating at full capacity; an economist's definition does not.

I-1 Short Clippings

A. Interpretation

1. The freeze shifts the supply curve to the left (damaged crops) and prices rise (inflated prices) to reach a new equilibrium.

3. Excess demand or deficient supply.

5. The key issue here is what one feels a good is worth and what someone will pay for the good. In the market the value of goods is determined by the interaction of supply and demand. In this clipping the vendor cannot drive up the market price unless someone is willing to pay this higher price.

7. Excess supply; the price is above the intersection of the supply and demand curves.

9. As in question 7. The rent cuts could be causing the falling vacancy rate.

11. a) There is excess supply; the price is above the intersection of the supply-and-demand curves.
    b) At current prices consumers will not buy all that is produced. The market will eliminate this resistance by pushing prices down until supply equals demand.

B. Shifting Demand And Supply Curves

1. The demand curve for nickel has shifted to the right causing prices to increase. How high the price increases will depend on the responsiveness of suppliers,

i.e., the steeper the supply curve the more the price rises. Since supplies are tight the supply curve will be steep resulting in the high prices.

3. The supply curve shifts to the left and prices rise.

5. a) A shift in the curve, since demand has increased at an unchanged price.
   b) A movement along, as a reaction to the price rise.

7. All tend to stabilize demand. Low inflation rates, stable interest rates and a stable economy reduce uncertainty and lead to less fluctuations in demand. Lower immigration levels mean less increase in demand as population growth is lower.

9. Improved grain production and increased acreage shift the supply curve to the right. Population increases shift the demand curve to the right. The supply-curve shift is more than the demand-curve shift resulting in lower prices.

11. A strike in B.C. would shift the supply curve for lumber to the left, increasing the price of lumber.

C. Substitution And Demand Changes

1. The high prices of beef and pork cause consumers to turn to substitutes, shifting the demand curves for these substitutes to the right, which tends to raise the prices of these substitutes.

3. The short supplies of pears and peaches will cause their prices to rise, turning consumers to substitutes such as cherries. Thus the demand curve for cherries will shift to the right, offsetting to some extent the effect on the price of cherries of the rightward shift of their supply curve.

5. The price of pork must be falling. As it falls, some consumers switch from beef to pork, shifting the demand curve for beef to the left. This puts downward pressure on the price of beef.

7. a) A shift in the supply curve, since it is stated that it is this fall in supply that has forced prices to fall rather than vice versa.
   b) Increased hog production will lower the price of pork, making pork more competitive with beef.

9. The rising cost of building materials will shift the supply curve for new homes to the left, and the price will rise. The demand for substitutes—older homes—will shift to the right causing their prices to increase.

D. Connected Markets

1. a) It's in equilibrium, with supply equal to demand.
   b) The shortage of grain means a leftward shift of the supply curve for grain, implying a rise in the price of grain. This increases the cost of producing eggs so the supply curve of eggs should shift to the left, leading to a rise in the price of eggs.

3. The drought has shifted the supply curve for feed to the left resulting in higher feed prices. The increased cost of feed has caused farmers to sell more hogs now than normal; this shifts the supply curve for pork to the right causing the fall in prices.

5. a) The demand curve for plywood should shift to the left and the supply curve for housing should shift to the right.
   b) Demand exceeds supply. Its price should rise in the short run because supply cannot increase to meet the demand. In the long run more plants will appear to supply the new product and the price should fall.

E. Miscellanea

1. In the short run the supply curve is quite steep, because it takes time for the higher price to lead to new copper mines. The demand curve is also steep, since it takes time for demanders to arrange for supplies of copper substitutes. In the long-run, the increased supply (new copper mines) and the lower demand (demand shifted to substitutes) lowers copper price.

3. Lumber may be purchased now rather than later, to avoid the anticipated price rise. This shifts the demand curve for lumber to the right and causes the price of lumber to rise.

5. The longer high prices exist relative to their substitutes the greater will be the substitution. Once users of copper make adjustments to other goods the more likely they are to continue their use, even if copper prices fall.

7. No. The rule of basic economics refers to increased supply leading to lower prices, *ceteris paribus*. In this case, the demand curve shifted to the right more than the supply curve did, so the expected price fall did not come about. The most likely cause of the demand increase is the higher quality of the 1970 vintage.

9. Low oil prices encourage consumers to increase their use and dependence on oil at the expense of alternate energy sources. Some high-cost suppliers of oil and alternate energy suppliers will leave the industry. This combination of increased demand and reduced supply will lead to higher prices in the future.

11. a) As farmers leave farming, the supply curve shifts left and prices rise to keep the remaining farmers in agriculture.
    b) Technical change and the move to larger, more efficient farms has caused a net rightward shift of the supply curve.

I-2A Natural Gas

1. If revenues fell by 0.5 per cent and sales rose by 6 per cent, then price must have fallen by 6.5 per cent. Elasticity is thus 6.0/6.5 = .92. Using the absolute numbers given in the article, however, the percentage quantity change is 6.27 and the percentage price change is 6.86 yielding an elasticity estimate of .91.

2. Suppose that in the absence of the price cut the demand for gas would have risen due to natural growth over time. Then the calculation in question 1 above uses a too-large quantity change attributable to the price fall and consequently overestimates elasticity.

3. Competition from substitutes such as oil may have forced them to lower the price of gas.

I-2 Short Clippings

A. Elastic vs Inelastic

1. Completely inelastic; a change in price causes no change in demand.

3. Elastic, since the revenues fell rather than rose when the price rise.

5. Inelastic. The lower price must have led to lower revenue for gas companies because consumers had more income to spend on other goods.

B. Calculating Elasticity

1. If demand in the U.S. would have increased by the same per cent as in Canada in the absence of a price increase, then the 20 per cent price increase is respon-

sible for decreasing demand by 7.3 per cent. This implies an elasticity of 7.3/20 = .365.

3. a) The old wage was 53/23 = $2.30. The new wage is 48/18 = $2.67. The percentage increase in the wage is .37/2.30 = 16 per cent. The percentage decrease in hours worked is 5/23 = 22 per cent. The elasticity is 22/16 = 1.4.
   b) The fact that total earnings fell reveals that the elasticity is greater than one and thus that the demand is elastic.

5. a) Assume for simplicity that the per cent revenue change is the sum of the per cent price change and the per cent quantity change. This implies an elasticity of 7/17.5 = .4. (Not using this simplification yields .34).
   b) Zero. For a 17.5 per cent price rise to increase revenues by 17.5 per cent, demand would have to remain unchanged.
   c) With a demand elasticity of .4, about 40 per cent of a price increase's effect on revenue will be wiped out by a fall in quantity. A price increase of about 30 per cent will therefore increase revenue by about 30-(30x.4) = 18 per cent.

7. Since the net quantity change is negative 0.03 per cent, the quantity change due to the price change is negative 2.03 per cent. Because the revenue change is plus 5 per cent, the associated price change must have been approximately 7.03 per cent. The elasticity is 2.03/7.03 = 0.29.

C. Miscellanea

1. No change in elasticity; the demand curve shifts due to the improvement in the economy.

3. a) The elasticity is 1/5 = .2.
   b) Since consumers are expecting further price falls they are not increasing their demand by as much as would otherwise be the case. This implies that the elasticity calculated above is too low.

5. The short-run elasticity is greater than the long-run elasticity because the drop in sales is less through time.

I-3 Short Clippings
A. Interpreting Cost Curves

1. a) Heating costs and unions would make the Canadian marginal and variable costs higher. The U.S. fixed costs would be higher because the operations are larger, but their average fixed costs would be lower. All three arguments would make the Canadian average-cost curve higher.
   b) Economies of scale.

3. The supply curve for tanker services will become horizontal at the scrap value price.

5. Our firms selling in the U.S. will receive less Canadian dollars for lumber due to the rise in the value of our currency. The squeeze on Canadian operations will be reflected by a lower marginal-revenue curve and therefore lower profits.

7. If animals are kept longer and fattened on feed they will become heavier and thus sell for more. But there are diminishing returns to fattening an animal—the extra feed consumed puts on less extra weight as the animal gets heavier (i.e., the marginal cost of putting on an extra pound of weight is rising). The farmer will fatten the animals to the point at which the rising marginal cost of putting on weight is just covered by the extra revenue from that extra weight. The fall in the price of feed lowers the marginal cost of fattening, so animals will be fattened a bit more before sale.

9. The wage subsidy will shift the marginal cost curve down and to the right. The firm will now be in a position where marginal revenue exceeds marginal cost. To reach the profit maximizing position the firm will increase output and employment.

11. a) Increasing output when price falls indicates a negatively-sloped supply curve.
    b) Profit-maximizing farmers would have been producing on the rising part of their marginal-cost curve where it equalled the marginal-revenue curve. Increasing output when the marginal-revenue curve is falling and marginal costs are rising would not make sense as it reduces profits.

B. Economies Of Scale

1. a) The marginal-cost curve and the average-total-cost curve are declining.
   b) If no collusion occurs, the price will be bid down until all but one newspaper is forced out of the market.
   c) Economies of scale, or natural monopoly.

3. a) The ATC is lower for 1100-passenger ships than 80-passenger ships.
   b) The demand for travel on small ships may be different or the demand may not be great enough to allow the larger ship to realize the economies of scale. Failing this, it may be that price covers AVC (perhaps because the opportunity cost of the small ship is zero) and the small ship will consequently operate in the short run.

5. The Canadian wine industry is still on the declining part of its long-run-cost curve.

C. Market Reactions

1. The cutting of power exports to the U.S. will cause the demand for coal to shift to the right and prices will rise. The increase in coal prices will cause greater output and employment as coal producers move up their MC curve and the market-supply curve.

3. a) A shift of the demand curve to the left by 10 per cent would occur.
   b) Since they wish to compensate fully for lost revenue, it must be the case that costs do not fall when demand falls: marginal cost is zero. As far as revenue is concerned, a 10 per cent rise in rates will offset the 10 per cent quantity fall only if the elasticity of demand is zero.

5. If a market is not profitable, then the consumers are not being served since it must be the case that the utility to consumers is not sufficient to make them pay a price high enough to justify production. Resources currently in this market will move elsewhere.

7. Setting price in Canada equal to marginal cost, with no premium to cover fixed costs.

9. The reason prices decline is because of surpluses or excess supply; the closing of mills is part of the market process of eliminating the excess supplies.

D. Opportunity Costs

1. The company is determining the opportunity cost of keeping its operations in their current use. If some of their operations can earn more elsewhere they will shift them over to that use.

3. a) Opportunity cost—they could earn this high interest rate by investing the cash in interest-earning assets.
   b) If demand and thus price increased, this company might be able to earn a

greater return by using the cash to increase production than by buying an interest-earning asset.

5. a) It shifts the cost curves upwards.
   b) Firms will leave the industry, shifting the supply curve to the left, until prices rise to a point where the remaining firms can cover their costs.

E. Shutting Down

1. When variable costs cannot be covered.

3. The price must be between their average-variable and average-total costs. In the short run they will at least contribute to some of their fixed costs and therefore minimize their losses.

5. a) Their current level of production is such that although fixed costs are not covered, variable costs are.
   b) Below marginal revenue.
   c) Demand for the output of the remaining mills should shift right as the sales of the closed-down mill are diverted to the others.

7. a) Profit.
   b) The entrepreneur requires recovery of opportunity costs—including a rate of return to compensate for the risk—of the resources allocated to the production of any product. What provides the incentive to the entrepreneur is the chance to earn at least the opportunity costs incurred. If this incentive is removed, entrepreneurs will not allocate more resources in response to market needs.

I-4A Newsprint

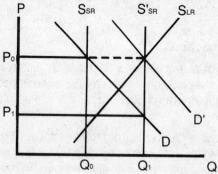

1. The current newsprint market can be represented by the intersection of the demand curve, D, and the short-run supply curve, S_{SR}, with the price at P_0. The long-run supply curve S_{LR}, indicates that at current prices the desired supply is Q_1. The current high demand is due to a shift of the demand curve because prices are rising.

2. The "$10 billion capital spending spree" is represented by an increase in capacity to Q_1, which at the current price is given by the long-run supply curve, S_{LR}. The new short-run supply curve with the increased capacity is now S'_{SR}.

3. The increase in capacity given the demand curve, D, will result in excess output of Q_1-Q_1 at price P_0. In order to eliminate this excess the market price will have to fall to P_1.

4. If the demand growth experienced in the past continues then the demand curve will shift to D' and prices will not fall.

5. When the prices of all goods are rising, but one good's price goes up more, then that good's price has gone up in real terms.

6. Lower prices. The supply will exceed demand as is indicated by the first part of the sentence.

7. The corn-hog cycle.

I-4 Short Clippings

1. a) The price of pork is relatively high.
   b) They should fall as the higher supply reaches the market.
   c) Production should shrink because of the relatively low price.

3. a) At a price greater than the equilibrium price and a quantity less than the equilibrium quantity.
   b) Pork supplies can be changed in a year, but beef supplies cannot be changed so quickly.
   c) Relatively high prices for beef.

5. The fall in price will discourage supply, so that by the time those waiting do decide to buy, the surge in demand that they represent in conjunction with the fall in supply could mean even higher prices.

7. Lower costs stimulate supply, but when the higher supply finally gets to the market it will create an excess supply at the current price and cause the price to fall.

9. a) The corn-hog cycle.
   b) It will cause prices and profits to fall.

11. The market for cars is now starting to move into the corn-hog cycle as supply has increased and demand is decreasing. These two factors will lead to falling prices for cars, which will result in losses for the industry, and in turn some car makers will leave the industry.

13. The industry in the short-run is experiencing high profits which is attracting additional capacity. The additional capacity will bring prices down such that the industry will be put into a loss position resulting in further adjustments.

I-5A Poultry and Eggs

1. The supply curve becomes a vertical line at Q_1 and price jumps up to its intersection with the demand curve.

2. At the higher price and lower quantity q_1 the farmer is probably operating above the ATC curve and so is making a higher income than before the marketing board when he or she was at the ATC curve at q_0. (We say "probably" because, if the ATC curve is very steep, it is possible that when the farmer is forced to cut output his average costs could increase by more than the increase in price.)

3. Consider first consuming the quantity Q_1. Under the marketing board, consumers pay an extra amount equal to the change in price times Q_1. Now consider the quantity Q_0-Q_1 which the consumers used to consume before the marketing board arrived. Their consumers' surplus from this purchase was given by the triangle above the original price between Q_1 and Q_0. These two items are added to obtain the total cost to the consumer.

4. Consumer surplus shrinks from the triangle above the original price to the triangle above the new price. Producers' surplus grows from the triangle below the original price to the area bounded by the new price, the vertical axis, the supply curve and the new quantity.

5. By the extent of the beyond-normal returns earned by a farmer because he or she has a quota.

6. This would increase supply and lower price, decreasing the incomes of the current quota holders.

7. The opportunity cost of the quota is what they could earn by selling the quota and investing the proceeds in interest-earning assets.

8. Without the quotas the corn-hog cycle would occur.

I-5 Short Clippings

A. Interpretation

1. Voluntary cutbacks in production.

3. a) If price is set above equilibrium, the consumer will not be willing to buy all that producers are willing to supply at that price.
   b) If the quota levels are set too high so that there is overproduction at the fixed price, the excess supply of eggs can be expensive to store.

5. a) A group of producers who enter into collusive agreements for their mutual benefit.
   b) Members of professional associations, e.g., doctors and lawyers.
   c) The professional associations do not want advertising because it would promote competition among its members, which could lead to price cutting.

7. The corn-hog-cycle nature of the oyster market.

B. Quota Values

1. The value price of an asset rises or falls due to changes in the income earned by that asset. For example, assume you have an asset which earns $1 and the market will pay you $10 for the asset, a 10 per cent rate of return. If the income earned by the asset rose to $2, ceteris paribus, the market would now be willing to pay $20 for the asset, a 10 per cent rate of return. In this clipping, subsidies and tax concessions raised the income of farms; therefore the price of farms is bid up.

3. a) The egg marketing board.
   b) The cost of buying a quota.
   c) The price of quotas is determined by the above-normal returns a quota permits its holder to reap.

5. The subsidies increase the income of property which increases the demand for it and in turn its price.

7. a) The above-normal rates of return (or income) to owners of quotas must be high.
   b) It would have to be at least $13,750, as $137,000 could earn that if invested at 10 per cent.

I-6A Minimum Wage

1. The minimum wage primarily affects the employment of the least skilled, heavily represented among the young. Overall unemployment in a province can be affected by many factors other than the minimum wage level (such as current world demand for that province's output). Thus to measure the impact of the minimum wage on unemployment it is necessary to correct for these other more general impacts on unemployment; examining the youth unemployment rate relative to the overall unemployment rate is one way of doing this. Another reason why high minimum wage rates could appear in provinces with low unemployment rates is that politicians may feel that their province can "afford" a high minimum wage if its unemployment rate is low.

2. a) Use the difference between the supply and demand curves at the minimum wage. Since it started at zero and is now positive, it has risen.
   b) Yes. Provinces with a relatively high minimum wage have an unduly high rate of unemployment compared to what would be expected given the province's overall rate of unemployment.
   c) Employment drops from its original level, at the intersection of the two curves, to the level given by the demand curve at the minimum wage.

d) The problem is that wage income rises because there is a higher wage, but falls since fewer are working. If the demand for labor curve is inelastic, total wage income will rise.

I-6 Short Clippings

A. Minimum Wages

1. a) The marginal revenue product (revenue contributed by hiring one more worker) is equal to the marginal cost (wages) of hiring that worker.
b) Let Figure I-6.2 represent the workers affected by the minimum wage. If minimum wages are raised from P_2 to P_1 then Q_2–Q_3 employees will be laid off.

3. The vast majority of the 38,000 workers were probably already earning more than $2. The relevant figure to cite is the percentage of those whose wage was below $2 and were laid off.

5. It is the largest group affected by minimum wages, because of their relative lack of skills. (Their marginal product is low.)

B. Rent Controls

1. Owners of rental accommodations have little incentive to maintain their buildings as they cannot pass on the costs to tenants. The building of new accommodations will also be reduced. The net effect of both, if rent controls are in place for a long time, is that city rental accommodations become older and more run down.

3. The subsidy to renters would shift the demand curve for rental accommodation to the right, and in the absence of rent controls would lead to a greater supply. (If rent controls were in force, this subsidy would further aggravate the current excess demand.)

5. When not enough rental accommodations are available the market price will rise which will induce more rental accommodations to be built.

7. The rise in demand for rental accomodation due to immigration and population growth will cause the price of rental accomodation to rise. In the long run this will induce landlords to increase the supply of rental accomodation, eventually lowering the price (rent), but in the short run tenants will be forced to pay higher rents.

9. a) The high vacancy rate indicates that there is no excess demand, so that removing the controls would not result in a price jump.
b) With low vacancy rates, the current price would be below the market-clearing price, whereas with high vacancy rates the current price would be close to the market-clearing price.

C. Miscellanea

1. If we price our oil too low, we will run out and have to turn to more expensive suppliers.

3. a) Under controls the price cannot be raised to clear the market.
b) The growing excess demand will eventually cause a black market to develop.

5. The government is fixing the price of soap and milk below the market-clearing price. Young people are reselling these items at the higher black-market clearing price. This is profitable for them to do if there is no opportunity cost for the time they spend waiting in line.

7. Both curves are highly elastic.

9. a) The price is too low.
   b) The money should be spent where society obtains more benefits than they are currently receiving.
   c) Charge a user fee for medical services; abolish rent controls.

I-7 Cross-Subsidization

1. Marginal revenue must exceed marginal cost on off-peak days; by lowering prices they will increase profits.

2. It probably *is* true for Voyageur. For each company, however, it is probably the case that marginal revenue being greater than marginal cost depends on the competition not matching a price reduction (i.e., most of the extra passengers gained by a price reduction come from the competition). Voyageur is objecting to the competition which would lower its profit on this run, a profit which needs to be greater than normal if Voyageur is to subsidize other routes.

3. Cross-subsidization implies that in market A the price is set higher than marginal cost, and the extra-normal profits from A are used to cover losses in market B, in which price is set lower than marginal cost. A is inefficient because too little is being produced; marginal benefit (measured by price) exceeds the cost to society (marginal cost). B is inefficient because too much is being produced; marginal benefit (price) is less than the cost to society (marginal cost).

4. By equating the price (marginal benefit) to the marginal cost of providing the service, society will receive the maximum benefit.

5. In Figure I-7.1(a), if we drop the price from where demand equals supply to P_0 then consumption will increase to Q_2 and society will bear a greater cost (the price on the supply curve corresponding to Q_2) for providing the last unit of service than the benefit it provides (measured by P_0). In Figure I-7.1(b), if we set the price at P_0 instead of where supply equals demand then Q_3 will be consumed and the cost of providing the last unit of service (the price on the supply curve corresponding to Q_3) will fall below the benefit it provides (measured by P_0).

6. A price is too high when it exceeds the marginal cost of providing it while price is too low when it is below the marginal cost of providing it. The price is just right when it equals marginal costs.

I-7 Short Clippings

1. . . . deal with averages.

3. When price is below ATC, losses are suffered. Shutting down completely means that the loss would equal the fixed costs. However if the price exceeds AVC, then by continuing to operate *some* of the fixed costs will be covered: The loss will be minimized.

5. Kentucky Fried Chicken should stay open to the point at which the extra cost of staying open an extra hour is equal to the extra revenue generated thereby. With the cost going up due to a rise in the minimum wage, it may be advantageous to KFC to cut back its hours.

7. Decisions on air cargo profitability should not be made on the basis of fully-allocated costs; marginal costs should be used.

9. a) The MC curve is rising. It is probably (but not necessarily) the case in this context that the AVC and ATC curves are also rising.
   b) Their total costs would be reduced, since their plant would not have to produce at high levels (where costs are higher).

I-8A Pollution

1. The cost curves would shift upwards. The resulting leftward shift in the supply

curve would raise the price and decrease the quantity. Pollution should be less at lower output levels.

2. It should charge an amount equal to the value of the disutility generated by the pollution.

3. Charging for the right to pollute allows the marketplace to continue to play an active role in the allocation and distribution of resources and permits the system to respond more quickly to any changes that might occur in technology or demand.

4. Polluters will bid the price of quotas to the point at which the consumer is just willing to pay a total price equal to the cost of production plus the cost of the quota. This will be equivalent to the result of question 2 only if the quantity (i.e., total quota) of pollution chosen happens to coincide with the quantity resulting from the application of the pricing rule of question 2.

I-8 Short Clippings

A. The Effects of Taxes and Subsidies

1. The supply curve shifts up two cents; the price rise depends on the slopes of the supply-and-demand curves.

3. It can happen if the demand curve is perfectly inelastic or the supply curve is perfectly elastic.

5. The subsidy will be "shared" by the tenant and the landlord depending on the slopes of the supply-and-demand curves, a result that can be seen by shifting the demand curve up by the subsidy.

7. a) The elasticity is 60/15 = 4. Note: Sales fell to 40 per cent of what they were before the price increase; therefore sales dropped 60 per cent.
   b) The extent to which it is passed on depends on the slopes of the supply-and-demand curves. For example, if demand is completely inelastic it will be entirely passed along.
   c) The removal of the tax shifts the supply curve to the right, reflecting an increase in supply at the current (after-tax) price. This causes a fall in equilibrium price, moving the consumer along the demand curve to a higher quantity; the producer moves along the new supply curve to a lower quantity, but this only partially offsets the initial increase in supply.

B. Pollution

1. a) The costs of reducing pollution are not noted.
   b) Quantification of the costs and benefits of reducing pollution versus reducing something else such as crime.

3. They could charge the exporters of power for the value of resources lost to alternative uses, and in the case of pollution they could tax their output to force them to internalize the pollution costs.

5. a) External diseconomies.
   b) By forcing the industry to internalize the cost of pollution they will find it in their best interests to lower the costs of preventing pollution.

7. Draw a supply-and-demand diagram for canned beverages, including in the supply curve the effect of a tax on cans accounting for environmental concern. These curves must intersect at a positive quantity; this reflects the consumer pressure.

C. Miscellanea

1. By holding the price of oil below the equilibrium price, demand in Canada continues to be high, forcing the importation of high-priced oil.

3. a) Removing rent controls allows rents to increase and thus provides the incentive for suppliers to provide more rental accommodation.
   b) The government is possibly trying to speed up the increase in the supply of rental housing while also preventing a steep rise in rents.

5. a) Over-use is use to the point at which the extra benefit of using an extra unit of a good or service is less than the cost of producing that extra unit.
   b) The economist would recommend a user fee set at the level of marginal cost. Each individual would then be motivated to not over-use the system.

7. The budget line AB would rotate around A to become steeper, reflecting the fact that because of the tax, necessities and food are now more costly. Full compensation will shift the new budget line out until it cuts through point C. The consumer will be able to now move to a higher level of satisfaction (a higher indifference curve) by consuming less necessities and food and more of other goods.

9. The demand and price paid for agricultural land is based on the profits earned from it. Decreases (increases) in subsidies decrease (increase) profits resulting in less (more) demand for land which in turn bids prices down (up).

I-9A Lumber

1. The tariff will cause the supply curve to shift up by 15 per cent with price settling where it intersects with the demand curve. The elasticity of the supply and demand curves will ultimately determine the price. For instance, the greater (lower) the elasticity of demand the lower (higher) the price increase.

2. If the demand curve is perfectly inelastic or the supply curve is perfectly elastic.

3. If the demand curve is perfectly elastic or the supply curve is perfectly inelastic.

4. The demand curve would have to be more elastic than the supply curve.

5. The reduced supply of Canadian lumber and its higher price will increase both the demand for U.S. lumber and its price. If they are perfect substitutes, Canadian mills will not be able to charge a higher price than U.S. mills because consumers will continue to demand more U.S. lumber until their prices are equal. If Canadian lumber is a higher quality product then prices may go above these levels.

I-9 Short Clippings

A. Tariffs

1. a) It is close to perfectly inelastic.
   b) The statement should read that U.S. consumers (rather than producers) would bear the full cost of the tariff.

3. The demand curve is highly inelastic. The demand curve would be nearly vertical so that when the price increases due to the countervailing duty, the decline in quantity is very small relative to the price increase.

5. a) The slopes of the supply-and-demand curves will decide how much of the tariff producers and consumers will absorb. For example, the more inelastic the demand curve, the greater (less) will be the amount of the tariff absorbed by consumers (producers).
   b) If the demand curve is completely elastic or the supply curve is completely inelastic.

B. Quotas

1. Quotas will shift the supply curve to the left, raising the price.

3. The only way the highly-paid autoworker could keep his or her job is for the quotas to force consumers to buy their products at high prices.

5. a) Although removing the import quotas would allow consumers to benefit, Canadian jobs would be lost.
    b) The savings could be measured by the change in the size of the consumers' surplus and the costs could be measured by summing up the lost wages and welfare costs. The time horizon is crucial. It may be that given time, the Canadian textile producers, if temporarily protected by quotas, could become efficient enough to compete with imports. This would shrink the benefits (savings) over time. On the other hand, it may be that the displaced textile workers could be retrained over time to obtain jobs in other industries. This would reduce the costs over time.

7. A more plausible explanation is that the quotas have restricted imports so that the imports that do come into Canada can be priced close to domestic textiles. Without the quotas, more imports would arrive to push the price of imports down.

9. The quota will reduce the supply of imports to America, causing their prices to rise. American cars are substitutes for import cars and therefore the demand curve for American cars will shift to the right and prices will increase.

11. a) See answer to question 9.
    b) The price will rise until 60,000 Canadians do not want to buy a Japanese car.

13. a) Canadian automobile dealers selling Japanese cars.
    b) For certain models, the demand will exceed the supply.
    c) They are few because only a small number of extra domestic cars are sold because of the quotas. They are expensive because all cars are sold at a higher price.

C. Free Trade

1. a) The law of comparative advantage.
    b) The price of goods and services will be lower.

3. The imposition of the duty will raise the price of goods and services to consumers but will not necessarily increase incomes (through more employment) enough to compensate for this increased cost. Countries affected by the tariff may retaliate against Canadian exports by imposing their own protective measures which may lead to more unemployment, lower incomes and less revenue to the government due to less income tax collected.

5. a) Protectionism leads to an increase in the price of the good being protected, in this case corn. This causes farmers to earn an above-normal return, causing them to bid up the price (and the rental price) of land.
    b) Farmers who own land when the protection is instigated will benefit; new farmers, and farmers presently renting land, receive no benefit.

7. a) The Mexicans would not grow strawberries if the proceeds from their sale did not more than compensate them for what they could have earned by growing corn. They should be able to buy more corn (with the income from growing strawberries) than they could have produced.
    b) B.C. consumers gain through lower prices; B.C. producers lose.

9. a) By making the inefficient firms unprofitable.
    b) The free market would create a lot of short-term unemployment which the government might be able to avoid by eliminating the inefficient industries gradually.

11. The total (i.e. including imports) supply curve shifts to the right, lowering price, but the U.S. domestic supply curve remains constant. As price falls there is a

movement along the domestic supply curve; some of the fall in quantity supplied comes about through the more inefficient of the U.S. firms closing down as the price falls.

13. Larger. Sales in the U.S. would have yielded more Canadian dollars at a lower exchange-rate value for the Canadian dollar.

15. Canadians could sell at whatever price was necessary to sell all their extra output in the U.S. and be assured they would not suffer a loss.

17. Flowers (workers) in inefficient industries will die (lose their jobs) as foreign competition puts these industries out of business; flowers in efficient industries will thrive as they expand into foreign markets.

I-10A Superstores

1. The new entry will shift the demand and marginal-revenue curves for the representative firm to the left.

2. The demand curve is highly elastic.

3. The answer to this question depends on the price at which the $1.5–$2 million sales are made. It probably refers to the left intersection of the firm's demand curve with its ATC curve.

4. The demand curve would be more inelastic and higher to reflect less substitutes. The independent operator will be able to charge a higher price for those things not offered by the superstores.

I-10 Short Clippings

1. The market has become more competitive.

3. Firms compete by offering more service rather than by offering a lower price. Their cost curves are identical; by offering extra service they hope to affect their demand curve.

5. a) There is a fixed amount of business and since there is an extra carrier to share this business there is less business for each carrier; requiring higher prices to cover their costs.
   b) Demand will be stimulated by the price fall, possibly by enough to allow the lower price to continue. If not, the less efficient carriers will go out of business and the new carrier will keep its price at least slightly below the old market price to prevent their re-entry.

7. In perfect competition all firms have the same cost curves and in equilibrium reach exactly the same price position.

9. a) Low. The move by De Beers to stockpile gems would be a rational move if the demand curve was inelastic. If they had put the gems on the market, prices would have fallen and De Beers total revenue and profits would have declined.
   b) The textbook equilibrium is when the market determines a price such that what is produced in a given time period is bought in that time period. If at a certain price more is produced than consumers will buy, prices will fall until what is produced will be sold. De Beers definition is to only supply enough output to meet demand at current prices.

11. a) Raising prices to a level higher than the minimum average total cost.
    b) A rise in price above the minimum level of the ATC curve will entice new entrants who will increase supply and drive the price back down to the lowest point on the ATC curve, where above-normal returns are zero.

I-11A Railroads

1. Off-line costs are fixed costs, while on-line costs are variable costs. The key question should be whether average variable costs are covered in this context, not average total costs. The economic viability of operating the branch lines should be determined on the basis of the marginal cost of operating them.

2. The marginal cost of operating the branch lines (i.e., the extra costs associated with their operation) should be compared with the marginal revenue associated with their operation (i.e., extra revenues, produced by their operation). Firms maximize profits by setting marginal cost equal to marginal revenue. Certainly they undertake any project for which marginal revenue exceeds marginal cost.

3. If the railbeds have been allowed to deteriorate to this extent, it's no wonder that the cost of operating the branch lines is high.

4. It shifts the bus company's demand curve left.

5. It is a subsidy to the bus companies, but a small one. The marginal cost to society of allowing bus companies to use the roads, which would be there and maintained anyway, is quite low.

6. a) Quantity Q_0 is determined by the intersection of MR and MC. Price P_0 is given by the D curve at this quantity.
   b) The cost curve would be quite high, above P_0 at Q_0. MC passes through its lowest point.
   c) Loss is shown by Q_0 times the difference between P_0 and the point on the average cost curve at Q_0.
   d) It should not affect price, since the company would still move to P_0 and Q_0, where losses are minimized. This result does not agree with Provincial Transport Enterprises Ltd.; they would object to this analysis, however, claiming that the railway cannot be considered a monopoly on some branch lines.

I-11 Short Clippings

1. Economic efficiency is enhanced when people pay an amount for a good or service that is equal to what it cost to produce an extra unit of that good or service. In this case the policy raises the price of electricity closer to the marginal cost of its production.

3. a) Marginal cost pricing.
   b) Since marginal costs are increasing as output goes up, this would mean that lavish users would pay higher prices than prudent users.

I-12A First Class Travel

1. The opportunity cost of the first-class seat must be taken into consideration. If tourist-class seats are consistently full, the opportunity cost of the first-class seat is the revenue it could be generating if converted to a tourist-class seat. When this loss of tourist-class revenue is taken into consideration, first-class service can be said not to pay if several first-class seats are consistently empty.

2. On some routes it may not be the case that some first-class seats are consistently empty so that on those routes they do pay their way. Or on some routes it may not be the case that tourist-class seats are consistently full, so that the opportunity cost of the first-class seats is very low.

3. A silly suggestion. Raising the first-class prices could easily mean that no one would travel first class, exacerbating rather than solving the problem. Scrapping them completely rules out the correct intermediate solution being pursued by the airline.

I-12 Short Clipping

1. It is price discrimination in that a different price is being charged to different consumers, but it is not discrimination in that the costs of serving these customers are not identical.

I-13A Freeway Gas Stations

1. The AC curve would be moved up, decreasing profit, but leaving unchanged the profit-maximizing price and quantity.

2. A percentage of the rectangle representing profit would have to be paid in rent. After-rent profit will still be maximized by maximizing before-rent profits, however, so price and quantity will not change.

3. The D curve rotates down around its intersection with the vertical axis by the amount of the tax. This also rotates MR down. Quantity will fall and price rise. Profits fall. Tax revenues are given as quantity times the gap between the demand curves at that quantity.

4. The rectangular area defining after-tax profits is the maximum bid that could be made.

5. Giving the answer to question 3 in reverse, the price should fall if the rent is reduced, but it will not fall by as much as the rent reduction. If the government wants a fall in price equal to the rent reduction, it will have to force that reduction.

I-13 Short Clippings

1. Draw a vertical line to the left of the intersection of the unconstrained annual supply and demand curves, representing the current, limited supply of cabs. The height to the supply curve measures the cost of supplying the annual service of the last cab, and the height to the demand curve measures the price that will be charged. Their difference reflects the annual return to running that cab, above the normal return built into the (unconstrained) supply curve. The capitalized value of this above-normal return (the dollar sum that would earn an annual interest payment equal to this return) is the price of a cab licence.

3. The increase in the number of taxis will reduce the profitability of existing medallion holders, which will reduce the value of their permits. Current holders of permits will be unhappy with this lower value, especially those who paid a higher price than will exist after the new permits are issued. Banks who lent money on the basis of the current value of permits may find that some lenders default on their loans.

5. a) They could set the price where the marginal cost curve cuts through the demand curve. In peak times when demand is high the landing fee would be raised, while in off-peak times the landing fee would be lowered.
   b) The profitability of both peak and off-peak times would increase, which will reduce the federal deficit.

II-1B High Wages

1. Keynesian theory suggests that unemployment can be eliminated by increasing aggregate demand, in particular by increasing government spending or reducing taxes, both of which would increase the government deficit.

2. The required diagram would measure the real wage on the vertical axis and the quantity of labor on the horizontal axis. The current real wage would be drawn in above the level at which supply and demand are equated. The level of employment is given by the demand curve at the current real wage, and the number of

people wanting a job at that real wage is given by the supply curve; the number of unemployed is given by the difference between them.

3. A major reason Canada's unemployment rate is so high is her high real-wage rate. The cure is to lower her real wage.

4. The right medicine is some policy to decrease the real wage, such as a rise in the price level without a corresponding rise in the money wage. The difficulties with this stem mainly from the fact that all sectors of society will try to resist any fall in their standard of living.

5. Each of these phenomena serves to make the Canadian real wage less able to flex in the downward direction. This explains why in the U.S. the real wage fell in response to unemployment whereas in Canada it did not.

II-1 Short Clippings

A. Labor-Force Growth

1. The former decreases unemployment because if people are not looking for work they are not counted as unemployed. The latter does the opposite.

3. In 1974 the labor force and employment grew at the same rate, so unemployment should have remained unchanged. In 1977 unemployment should have risen as the labor force grew by more than employment. In 1979 employment grew faster than the labor force so that unemployment should have fallen.

5. The number of people wanting jobs (the supply of labor) must have increased by more than the number of jobs created (the demand for labor).

7. a) A drop in the unemployment rate.
   b) The unemployment rate drops only if the number of people entering the labor market is less than the number of new jobs.

9. Women invading the labor force caused the annual increase in labor supply to be greater than the annual increase in labor demand (jobs created), increasing the rate of unemployment.

B. Participation Rate

1. a) The percentage of females wanting a job, usually with reference to all females between the ages of 15 and 65.
   b) No. We don't know how many of the extra workers in each category (male and female) were able to get jobs.

3. A rising participation rate means that more people want jobs, increasing the unemployment rate.

5. . . . less people were looking for work, moderating the rise in the unemployment rate.

C. Discouraged and Encouraged Workers

1. As the economy picks up, many people previously too discouraged to look for work may resume their search and will now be counted as unemployed.

3. The unemployment rate may have fallen because many of the unemployed became too discouraged to look for work and thus are no longer counted as unemployed.

5. The marginally-attached worker enters the labor force when the economy is booming, preventing the unemployment rate from falling as much. When the economy moves into recession, these people leave the labor force, preventing the unemployment rate from rising as much as it otherwise would.

D. Unemployment Insurance

1. a) Some people arrange to "lose" their jobs so they can "vacation" on unemployment insurance.
   b) Because unemployment insurance is so generous, people will take longer to find a new job—they can afford to be fussier.
   c) People on unemployment insurance maintain their demand for goods and services, but do not contribute to supply.

3. They both increase the unemployment rate.

5. a) Seasonal industries are subsidized because an employer will not have to pay as much to entice someone to work in a seasonal industry as he or she would if the employee did not have available unemployment insurance during the off-season.
   b) With experience rating, seasonal-industry employers would have to pay more into the unemployment insurance fund. This should increase their cost of labor, which would decrease their demand for labor and thus decrease the wage paid.

E. Miscellanea

1. a) Workers trained and experienced in working for a particular employer are valuable to that employer; in laying them off the employer runs the risk that they will not be available once the economy recovers.
   b) The lag is due to the fact that output can be increased by utilizing the hoarded labor instead of hiring extra labor.

3. Compared to the time of the Great Depression, more of today's unemployed are secondary-income earners whose job loss does not necessarily leave their family without adequate income.

5. a) The percentage of people (in a certain age and sex category) who are employed.
   b) Men in this category have less experience and skills than older men, so their employment would be more sensitive to cyclical change.
   c) One is not 100 per cent minus the other, since those voluntarily unemployed are omitted when calculating the unemployment rate. Fluctuations in the number of voluntarily unemployed affect the unemployment rate but not the employment rate, rendering the latter a more reliable indicator of cyclical change.
   d) The economy is moving from the low of the cycle to the high, since the employment rate has jumped.

7. a) Central bankers, through monetary policy, often attack inflation through the creation of high unemployment.
   b) Wages higher than the wages at which those without jobs (the outsiders) are willing to work.

II-2A Uncertain Saving

1. By increasing spending, consumers are increasing aggregate demand and thereby stimulating the economy to grow to a higher level.

2. The 9 per cent saving rate cited is saving as a percentage of personal disposable income, so the average propensity to consume must be 0.91.

3. Suppose inflation has recently been 20 per cent, with an interest rate of 24 per cent, implying a real interest rate of 4 per cent. Now suppose inflation falls to 10 per cent and the interest rate falls to 15 per cent, so that the real interest rate is 5 per cent. The real interest rate (the return from saving, after accounting for the

erosion of the savings value due to inflation) has risen to 5 per cent, so some people may be enticed to save more. But others may see only the fall in the nominal rate of interest and react by increasing spending.

4. Smaller. At higher income levels people can utilize the tax deductions which cause their saving to increase, and thus their consumption decreases.

5. It decreases consumption, decreasing aggregate demand and thereby also decreasing national income.

II-2   Short Clippings

A.   Savings and the Savings Rate

1. a) A rise in the APC.
   b) The initial use of "savings" refers to current saving out of income, whereas its later use refers to funds that have accumulated due to past saving.

3. a) A rise in the APC to 0.925.
   b) The aggregate demand line on a 45° line diagram will shift up, leading to an increase in income. Alternatively, the aggregate demand curve on an aggregate demand-aggregate supply diagram shifts up, leading to the same conclusion.

5. No. The figure refers to funds that have accumulated due to past saving.

B.   Consumption Behaviour

1. Interest rates and consumer attitudes.

3. a) A change in consumption due to a change in the wealth of the populace.
   b) A stock market crash would not affect the wealth of a majority of Canadians, and so no substantive wealth effect would be expected.

5. Since sales taxes are to rise, people will now find saving relatively more attractive than before, so we would expect saving to increase a bit more than would otherwise be expected.

C.   Influence on Income

1. a) The APC rises to 0.90.
   b) The higher savings rate would decrease consumption, decreasing aggregate demand and thereby exacerbating the recession.

3. The decrease in consumer spending has decreased aggregate demand which has lowered income and thereby decreased income-tax receipts.

5. By increasing consumption, aggregate demand is increased which stimulates the economy to move to a higher level of income.

7. The wealth effect.

II-3A   Recovery Stronger

1. The multiplier process.

2. During a recession, firms tend to "hoard" labor—keep on more labor than is necessary to produce the output demanded—so that they will not run the risk of losing skilled workers experienced in working in their particular firm. When recovery comes this hoarded labor can be used to increase output without hiring extra workers, so that the measured productivity of these workers increases dramatically.

3. To reduce the rate of inventory liquidation it is necessary to increase output to replenish inventories at a faster rate.

4. High interest rates decrease investment and other demand, and a faltering global economy would decrease exports; both decrease aggregate demand for domestically-produced goods and services and thus act to dampen the domestic economy.

5. If consumer confidence is high, consumer spending should hold up and thus maintain the recovery.

II-3 Short Clippings

A. Interpreting Inventory Changes

1. It is probably due to a fall in aggregate demand, which will move the economy into a recession.

3. The rise in inventories suggests there will be a fall in aggregate demand and that the economy may move into recession.

5. Firms will cut back on production to try to bring their inventories down.

7. Looking not at production, but at aggregate demand.

9. Strong consumer spending will stimulate producers to increase output; a buildup in inventories will induce them to cut back on output.

B. Forecasting with Inventories

1. Increases in consumer spending directly increase aggregate demand and thus stimulate the economy. Reductions in inventories imply that output will soon have to be increased to build inventories back up to their desired level or to prevent them from falling below this desired level.

3. Firms will rid themselves of the extra inventories by cutting back on production.

5. A cutback in production designed to make inventory levels fall back to normal levels. The drop in production reduces current output—"this quarter's real gross domestic product."

7 The inventory change (falling) tells us that aggregate demand is greater than production, suggesting that soon producers will react by increasing output. The inventory level (low) tells us that firms will not wish to allow inventories to continue to fall, so they will increase production.

C. Determinants of Investment

1. High levels of capacity utilization and high profits.

3. Lower wage increases imply that operating new machinery is cheaper, and therefore more profitable. A lower Canadian dollar implies that equipment-producing exports will be more profitable. Lower government growth implies that corporate taxes are less likely to rise, making investment projects more profitable. Higher rates of capacity utilization lead business to expect that extra capacity will be needed in the future, and that therefore investment in extra capacity will be a profitable move.

5. a) The level of income.
   b) Capital spending is a component of aggregate demand for goods and services. If it falls, income will fall according to the usual Keynesian mechanism.

D. Investment Impact and Timing

1. The fall in investment creates a fall in aggregate demand which will move the economy into a recession.

3. When the economy is operating at a high capacity utilization rate, new invest-

ment will be undertaken to allow business to increase output. This increases aggregate demand and stimulates the economy.

5. Decisions to undertake capital spending are usually made during a boom. By the time these decisions are translated into actual spending, the economy is usually past the boom and has moved into recession.

II-4 Short Clippings

A. Government Spending

1. Through the multiplier effects of an increase in government spending.

3. Keynesians believe that government policy (usually fiscal policy) is needed to move an economy out of a recession because any automatic forces that might right the economy operate too slowly.

5. a) Fiscal policy.
   b) The U.S. economy will be stimulated by this fiscal policy; the resulting increase in U.S. income will increase U.S. demand for Canadian exports, increasing aggregate demand in Canada.

7. Without creating inflation, or crowding out.

B. The Multiplier

1. a) Fiscal policy.
   b) The multiplier process.

3. a) The ultimate increase in the number of jobs per dollar increase in spending.
   b) Either 162 or 202 divided by 1 million.

5. a) The multiplier refers to the dollar value of an increase in output due to a $1 increase in government spending, whereas the real multiplier refers to the real (physical) increase in output due to a $1 increase in government spending.
   b) The real multiplier would be smaller because some of the stimulus due to the higher government spending will manifest itself in higher prices.
   c) Comparing the Q values associated with the new and old equilibrium positions yields the real multiplier. Comparing the rectangles formed by the origin and the equilibrium positions yields the multiplier.
   d) If the aggregate supply curve is horizontal.

C. Tax Policy

1. a) The standard multiplier mechanism.
   b) . . . inflation.

3. Without indexing, as nominal incomes rise (with no change in real income because of inflation) taxpayers will move into higher tax brackets. Thus they will pay a greater fraction of their income in taxes, even though there has been no increase in real income.

5. An increase in government spending of $1 increases aggregate demand by $1, but a decrease in taxes of $1 only increases aggregate demand by the MPC times that dollar since some of that extra dollar of disposable income is saved. (Also, some is spent on incomes.)

D. Budget Deficits

1. The standard multiplier mechanism. If income increases by enough, in response to this stimulus, tax revenues might rise by enough to liquidate the debt.

3. To cut the deficit either taxes must be raised or government spending must be cut, both of which lower aggregate demand and dampen economic activity through the multiplier process.

5. By claiming that such deficits are necessary to keep the economy at full employment.

7. The deficit could come about by a fall in tax receipts due to a downturn in the economy rather than by a dose of fiscal policy.

E. Miscellanea

1. The government borrowing will raise the interest rate, which will in turn cause aggregate demand of several types to fall.

3. . . . save more now so as to be able to pay the higher future taxes without a marked disruption to their consumption level, the resulting fall in consumer demand will "crowd out" the stimulation of the deficit.

II-5A T-Bill Auction

1. The average bid price must have fallen—if less is paid for the bill, then when it is redeemed at its face value the yield is greater.

2. By adding 0.25 per cent to the average yield.

3. By paying 97.300 one obtains $100.00 in 92 days. Thus $2.70 must correspond to an annual yield of 11.13% on $97.30 over 92 days.

4. The Bank would have bid for more of the offering at higher prices.

II-5 Short Clippings

A. Changing the Money Supply

1. Banks need only hold a fraction of their deposits in the form of reserves. Thus an extra dollar of reserves will lead to a multiplied increase in the supply of money (via what is called the money multiplier).

3. Credit expansion is an increase in the money supply, which is apparently growing at an excessive rate, promoting inflation.

5. a) The chartered banks would use these deposits as reserves on which they can base new loans, thereby increasing the money supply. For this to be the case some of the bonds must have been sold to the Bank of Canada, creating new reserves.
   b) Because the government wants to spend the proceeds.

B. The Bank Rate

1. a) He could claim that it is the market, not his policy, that has increased the interest rate.
   b) The Bank of Canada influences the market by buying more or less of the bills at the weekly auction.

3. The bank participates in the auction for 91-day treasury notes. If, for example, it buys an extra amount of these bills, the price of the bills will be bid up and the treasury bill rate (and thus the bank rate) will fall.

5. a) The interest rate will be prevented from rising.
   b) The money supply will increase by the half billion dollars times the money multiplier.

C. Measuring Money

1. People may be keeping as much of their transactions money as possible in interest-bearing forms. These forms are included in M1A but not in M1.

3. Because people's money-demand behavior will be affected and so their past behavior cannot be used as a guide for policy.

5. . . . the level of national income.

7. The size of the money holdings in that kind of account can no longer be used as a guide to people's spending behavior, and therefore will not be as strongly related to the level of economic activity.

9. a) Money that is used for the purpose of lubricating spending.
   b) So they can maximize the interest return on their cash holdings.
   c) By encouraging this activity the banks could increase their loans by more than would otherwise be the case (because of the lower reserve requirement on this type of account), increasing the money supply.

D. A Monetary Rule

1. A monetary rule.

3. a) Expanding the money supply at a fixed rate.
   b) If inflation proceeds at a rate higher than that which can be supported by the rate of growth of the money supply, interest rates will rise and aggregate demand will fall, choking off the inflationary pressures. This rule is being abandoned because it has not been possible to find a strong enough relationship between economic activity and a measure of the money supply to justify using some measure of the money supply as the appropriate target.
   c) Probably a subjective judgement about interest rates and inflation rates, and where they are headed.

5. If the economy overheats, the fixed money supply growth will be insufficient to meet the demand for money and the interest rate will rise.

7. a) The rate of inflation will equal the difference between the rate of growth of the money supply and the rate of real growth of the economy.
   b) As income grows, the demand for money grows. If the supply of money grows by more than this, there will be excessive money—people will not wish to hold this money in the form of money. One way to get rid of these excessive holdings of money is to spend them on goods and services, bidding up their prices. A second way is to buy financial assets, bidding up their prices which causes interest rates to fall. This increases aggregate demand for goods and services, causing the price level to rise.
   c) In the short run, major shocks to the price level, such as the oil crisis, can cause inflation to deviate temporarily from what the monetarist theory says it should be over the long run.

E. Monetary Policy

1. Monetary policy could also be defined as the control of the money supply. Since the interest rate is controlled by the Bank by effecting changes in the money supply, it is more appropriate to define monetary policy in terms of control of the money supply.

3. A continually-increasing money supply.

5. a) Not increasing the supply of money to match increases in the demand money at the prevailing interest rate
   b) Increasing government spending or decreasing taxes
   c) Use of monetary or fiscal policy to start the multiplier process in motion, as priming a pump is required to get it going

7. Lower interest rates stimulate the economy, increasing income and thus increasing the demand for money. This increase in the demand for money will tend to push interest rates back up, dampening the economy—something the govern-

ment may not wish. To prevent this, the money supply will be increased to meet the higher demand for money at the prevailing interest rate.

F. Regional Policy

1. . . . inflationary pressures.

3. . . . the federal government, which controls fiscal policy.

5. By using fiscal policy that increases spending in weak regions and decreases spending in overheating regions.

7. By raising interest rates so high that they depress demand for housing.

II-6A Measuring Inflation

1. A cost-of-living index incorporates into its calculation the fact that relative quantities of goods and services purchased change as relative prices change. A price index is calculated holding constant the relative quantities of items purchased.

2. Overstate. As the price of a particular good rises, people will tend to buy less of that good, implying that the impact of that price rise on our cost of living is not as great as if we had continued to buy the same quantity of that good. Since the price index assumes that we do not change the quantity of that good that we purchase, it overstates the change in the cost of living.

3. The percentage rise in the cost of the basket is (952 - 680)/680 = 40 per cent, so the price index will rise from 100 to 140. The average annual rate is 5 per cent.

4. The prices of goods on which we spend a lot are given heavier weight than the prices of goods on which we spend very little. The weights used are the fractions of a typical consumers budget spent on the different goods and services. They are obtained through the family-spending survey.

II-6 Short Clippings

A. Measuring Inflation

1. The higher cost of food has caused people to spend less on food, relative to other things, than they did before the price rise. Consequently its weight in the price index should fall.

3. a) The price of a good embodies indirect taxes, so whenever indirect taxes increase, the price rises and affects the inflation measure.
   b) No effect, except insofar as they dampen aggregate demand and thus eventually relieve pressure on prices.

B. Causes of Inflation

1. The usual argument is that higher government spending increases aggregate demand and exacerbates inflation through demand-pull channels. The argument in this clipping is more accurately labelled cost-push.

3. a) Monetary and fiscal policy action.
   b) These policies increase demand beyond the point at which supply can meet this demand; hence, prices rise.

5. . . . if those experiencing a fall in their cost of living insist on avoiding this fall through cost-push forces, and these cost-push forces are validated by the monetary authorities.

C. The Role of Money

1. If output (income) increases by, say, 5 per cent, then the demand for money will increase by 5 per cent. If the supply of money increases by more than 5 per cent, then the extra money will be spent, bidding up prices.

3. See the answer to question 1 above.

5. a) The rate would be equal to the rate of real growth of income.
   b) In the short run it is possible to have the rate of inflation unaffected by a change in the rate of growth of the money supply. Also, it may be the case that institutional changes in the banking industry are changing the roles played by different monetary aggregates in effecting spending, as well as making them more efficient. M1 may therefore be the wrong monetary aggregate on which to be focussing attention.

D. Unemployment Implications

1. If wages and prices continue to rise, but the money supply does not increase at the same rate, the real supply of money will fall and cause the interest rate to rise. This will dampen aggregate demand and push the economy into unemployment.

3. The government protection consists of employing monetary and fiscal policies to ensure that demand and employment remain strong. The results would be a fall in sales and consequent unemployment.

E. Miscellanea

1. a) Fixing the rate of growth of the money supply.
   b) Inflation will increase the demand for money. If the supply of money does not increase by an equivalent amount, the interest rate will rise, dampening the inflationary forces.
   c) Interest rates will have to be allowed to fluctuate as in the answer to part b).

3. High money supply growth creates inflation only if it grows faster than the demand for money.

5. a) At low interest rates and an inflation rate of 8 or 9 per cent, real estate speculators couldn't help making money.
   b) To achieve low interest rates in the face of a high inflation rate (which itself makes the nominal interest rate high) it would be necessary to accelerate the rate at which the money supply is increasing, to outstrip expectations. This will accelerate the inflation.

II-7A Challenging Conventional Wisdom

1. As income grows, the demand for money grows. If the supply of money grows by less, there will be an excess demand for money. Those wanting the money will bid up the interest rate to obtain it.

2. The interest rate observed in the market incorporates a premium for the expected rate of inflation. If market participants believe that an increase in the rate of growth of the money supply will push inflation up, as the monetarists claim, their expected rate of inflation will rise, causing the interest rate to rise.

3. Most economists would agree with Miller for the short run and Herskowitz for the long run. In the short run interest rates would rise if people do not immediately change their inflationary expectations when the rate of growth of the money supply is slowed.

II-7 Short Clippings

A. Bond Prices and Yields

1. a) The rise in the price of bonds, caused by the fall in the interest rate.
   b) Short-term bonds will soon be redeemed at their issue price, so the price of a short-term bond will not differ much from this issue price.

3. . . . capital gains if the interest rate falls and the price of the bond increases.

5. If inflation increases, people's expectations of inflation increase causing the interest rate to rise. This causes the price of bonds to fall.

7. The smaller-than-expected decrease in the money supply must have caused people to increase their expectations of inflation which in turn led to a rise in the interest rate. This implies a fall in bond prices.

B. Bond Markets

1. a) A fall in the interest rate will cause a rise in the price of bonds.
   b) It could lead to inflation and a rise in the nominal interest rate.

3. a) Rising interest rates are lowering the price at which they can sell these bills.
   b) The shorter-term bills do not vary so much in price.
   c) The rising loan demand has pushed up the interest rates.

5. If Washington balances its budget, the federal government will not be trying to borrow so much, so interest rates should fall. If OPEC abandons support for world oil prices the world price of oil should fall, alleviating inflationary pressures. This reduces expected inflation and thus causes the interest rate to fall. The fall in the interest rate, from both these sources, causes the price of bonds to rise.

7. a) It is bad in that the economy is weaker, but good for those working in the bond market because the price of bonds rose.
   b) Economic weakness should decrease the demand for money which should put downward pressure on interest rates, and the monetary authorities may also take steps to stimulate the economy by lowering interest rates. The lower interest rates imply a rise in bond prices.

9. a) The money supply surge will cause dramatic changes in people's expectations of future inflation, which will change the interest rate, in turn affecting the price of long-term bonds. If the price of long-term bonds becomes volatile because of this, people will no longer be willing to invest in them.
   b) Short-term bonds will soon be redeemed at their face value, so they will not experience much change in their price.

C. Real vs Nominal

1. The real interest rate is the nominal interest rate less the expected rate of inflation. If we measure the expected rate of inflation by the per cent of the CPI rise, then real rates are 8% - 4% = 4%.

3. It suggests that high nominal interest rates, high because of a high expected rate of inflation, do not have much impact on economic behavior, implying that it must be the real interest rate that is the primary determinant of economic behavior.

5. The nominal interest rate is high, but the real interest rate, which determines borrowing behavior, is not.

D. Inflation and Interest Rates

1. Rise. A rise in expected inflation increases the nominal interest rate.

3. a) If inflation rises, expected inflation will rise and the nominal interest rate will rise along with it.
   b) They will suffer a capital loss because the price of the bonds will fall as the interest rate rises.

5. a) There are two impacts of higher interest rates on inflation. The higher rate

raises the cost of doing business and thus increases prices, promoting inflation. The higher rate also dampens aggregate demand, reducing inflationary pressure.

    b) Higher expectations of inflation do increase the interest rate, so a jump in the interest rate could easily foretell an increase in inflation if people's expectations are correct.

7. The first half is talking about the fact that high *real* rates of interest can dampen aggregate demand and thus reduce inflationary pressure. The second half is talking of the rise in the nominal interest rate to match the rise in inflation.

E. Forecasting

1. A recession means a lower income level and thus a fall in the demand for money, tending to decrease the interest rate. A higher inflation means higher expectations of inflation, increasing the nominal interest rate.

3. Growth in the U.S. will raise the demand for money, putting upward pressure on interest rates. The high rate of growth of the money supply will increase expected inflation, raising the nominal interest rate.

5. A strong demand for new treasury issues suggests that the interest rate on these issues is a little high (i.e., demand exceeds supply at that interest rate) and it should fall in the future.

F. Lowering Interest Rates

1. a) The extra money supply causes the supply of money to exceed demand for money, so its price, the interest rate, should fall.
    b) The higher rate of growth of the money supply will soon raise expectations of inflation which will cause the interest rate to rise.

3. The first sentence is backward—the key to lower interest rates is containing inflation. The current high interest rates are designed to fight inflation; when inflation falls, the interest rate should fall as well.

5. Lowering inflation will lower people's expectations of inflation, which in turn will lower the interest rate.

7. If the Bank tried to lower short-term rates by too much, the increase in the money supply required to accomplish this would raise expectations of inflation, thus raising the interest rate.

G. Policy

1. High interest rates could indicate that monetary policy was tight, in that the high rates could be the result of a restrictive, inflation-fighting monetary policy. However, high interest rates could also result from an easy monetary policy that has induced high inflation. The high inflation raises inflation expectations which creates the high interest rates.

3. The Bank will raise interest rates to decrease aggregate demand to relieve inflationary pressures, i.e., as an inflation-fighting measure. They also will be forced to raise interest rates because as inflation moves up, expectations of inflation rise, and if the Bank does not raise interest rates it will find itself increasing the money supply at an ever-increasing rate to prevent a rise in the interest rate due to market forces. This would make the inflation worse.

5. If the interest rate is to be held constant, as the clipping implies, any change in the demand for money must be met by a change in the supply of money.

7. Although running out of money does cause high interest rates, so does too

much money, through its impact on expected inflation. The high interest rates must be high real-interest rates if they are to have any success in stopping the economy.

II-8B Big Mac Index

1. The purchasing power parity theorem.

2. It should be the same everywhere.

3. If something costs more in country A, then it will be imported from the country in which it is cheaper, forcing the domestic price to fall.

4. The Big Mac cannot be exported to other countries, since it is perishable. Further, local labor costs play a large role in the price of a Big Mac; so as long as labor is not free to flow across borders, this component of the price difference will remain.

5. A Big Mac costs $2.19 in the U.S. and $1.63 in Canada, both in U.S. dollars. PPP says they should cost the same; this would be the case if the Canadian dollar were 25% higher (2.19 - 1.63)/2.19 = .256.

6. The author is suggesting that government leaders do not take appropriate policy action when their currencies are overvalued.

II-8 Short Clippings

A. Trade Deficits and the Current Account

1. An overvalued C$ would shift the aggregate demand curve to the left, through its negative impact on the demand for Canadian exports and the shifting of demand from domestically-produced goods to imports.

3. a) A rise in the value of the U.S. dollar.
   b) Demand for domestically-produced goods and services falls and through the multiplier process dampens U.S. growth.

5. a) It causes demand for U.S. exports and demand for import-competing goods to fall, and thus causes a negative aggregate demand shock.
   b) Adding up all spending in the economy during the year. The excess is subtracted from the sum of consumption, investment and government spending. It is subtracted because the figures for consumption, investment and government spending include spending on imports which have not been produced domestically.

7. a) . . . a deficit on the current account.
   b) If the government does not participate in the foreign exchange market, supply and demand in the foreign exchange market can only come about by this matching.

9. The budget deficit requires that the U.S. government borrow a lot, pushing up the U.S. real interest rate. This causes capital inflows (which could be caused directly if the government borrows from foreigners). The capital inflows raise the value of the U.S. dollar which in turn causes U.S. exports to fall and imports to rise, creating a trade deficit.

B. Determinants of the Exchange Rate

1. Chevron will take the Reichmanns' Canadian dollars and buy U.S. dollars with them. This increases the supply of Canadian dollars on the foreign exchange market, putting downward pressure on the value of the Canadian dollar.

3. a) A strong economy usually implies higher income and thus higher demand for imports, putting downward pressure on the C$.

b) The reason the Canadian economy is strong may be that exports are high.

5. The fall in the current account implies a movement towards a balance of payments surplus which will imply a rise in the value of the C$. If the C$ is expected to rise, professional currency traders will buy it in hopes of making a capital gain.

7. a) A current account surplus, if it corresponds to a balance of payments surplus, results in excess demand for the C$ on the foreign exchange market and thus a rise in its value.
   b) The declines in the C$ have led to the increases in the current account surpluses, rather than the other way around.
   c) To abstract from the different levels of inflation in the two countries.

C. Fixing the Exchange Rate

1. As the economy tries to recover, the higher level of income would raise imports and create a balance of payments deficit. With the exchange rate fixed, the money supply would automatically contract, raising the interest rate and dampening the recovery.

3. a) See the answer to question 7 below.
   b) The comment in the last sentence suggests that U.S. inflation is less than ours. This implies that fixing the exchange rate would force our money supply to grow at a slower rate.

5. The problem of fighting inflation with a fixed exchange rate.

7. Higher inflation in the U.S. with a fixed exchange rate meant that its trading partners experienced balance of payments surpluses. Under a fixed exchange rate this led to automatic increases in the money supplies of these countries (i.e., as the U.S. dollars corresponding to the balance of payments surplus are traded in at the fixed exchange rate for domestic currency), bumping their inflation rate to match that of the U.S.

9. By fixing the exchange rate in terms of the U.S. dollar, we have been forced to adopt a monetary policy identical to that of the U.S.

D. Managing the Exchange Rate

1. . . . prevent the Canadian dollar from falling.

3. . . . prevent the Canadian dollar from rising.

5. a) The high value of the Canadian dollar means that exporters do not receive as many Canadian dollars for their foreign currency revenues.
   b) . . . selling Canadian dollars.

7. The capacity constraints argument is that although a fall in the exchange rate will increase demand for Canadian exports, because the export industries are operating at full capacity this will result in a rise in the price of Canadian goods rather than an increase in output, exacerbating inflation. Since Canada is not operating at full capacity, the proposed policy would seem suitable.

9. a) The dollar is falling in value.
   b) Using foreign reserve holdings (i.e., selling foreign reserves) to buy Canadian dollars on the foreign exchange market, thereby putting upward pressure on its price.
   c) The higher Canadian interest rate will attract capital inflows, putting upward pressure on the Canadian dollar.

11. a) Market intervention means buying or selling the Canadian dollar, thereby affecting its price. Monetary policy can change Canada's interest rate, making it more or less attractive for foreigners to buy Canadian dollars to invest

in Canadian bonds. This change in the demand for the Canadian dollar affects its price.

b) A country which intervenes in the market to affect the price of its currency. This may be done to push the price in a desired direction or, more acceptably, to prevent unnecessary fluctuations in the price.

13. a) Fall. The Bank is buying Canadian dollars, increasing the demand to prevent its price from falling.

b) The Bank would run out of foreign exchange reserves.

15. The higher U.S. interest rates will cause a Canadian balance of payments deficit because capital will flow out of Canada to seek the higher return in the U.S. If the rise in the Canadian interest rate and the fall in the exchange rate is not sufficient to eliminate this balance of payments deficit, Canada will experience a loss of international reserves.

17. To stem the rise in the value of the dollar, the government would be selling Canadian dollars on the foreign exchange market. To get the dollars they could just print them, but that would create an undesirable rise in the Canadian money supply. To prevent this they would need to get the dollars by selling bonds on the domestic financial markets.

E. Purchasing Power Parity

1. Purchasing power parity.

3. First, it is the *annual* change in the inflation rate that is relevant. Second, the reference in the last sentence to the *change* in the inflation rate does not make clear that it is implicitly being assumed that inflation has changed in our country but not in the other country.

5. The benefit of the lower dollar has been offset by the higher rate of inflation in Canada.

7. The second paragraph undertakes its calculation by simply converting country A's income into U.S. dollars using the prevailing exchange rate. The first paragraph instead asks how much of a standard bundle of goods and services bought in the U.S., using U.S. income, could be purchased in country A using country A's income.

9. a) Canada usually has a current account deficit, offset by a capital account surplus. Forcing the current account to be in balance would therefore tend to create a balance of payments surplus in Canada.

b) Below, since the exchange rate would have to fall to eliminate the current account deficit.

F. Inflation and the Exchange Rate

1. The higher inflation in Canada would make her exports more expensive and her imports cheaper, creating a balance of payments deficit. A continual fall in the exchange rate would prevent these price differences from arising.

3. a) If Canadian prices are rising faster than U.S. prices, and if the exchange rate is fixed, the cost to Americans of Canada's exports rises and Canada's exports will fall. This fall in aggregate demand leads to unemployment in Canada.

b) By letting the exchange rate fall.

5. a) Yes, if the exchange rate is fixed. The price of imports will remain steady as domestic prices inflate (assuming our inflation is not matched by foreign countries), so imports will rise. Our exports will become more expensive to foreigners, so exports will fall. If the exchange rate is flexible, the flexing of

the exchange rate will offset the inflation differential, and there should be no pressure for imports to exceed exports.

b) Interest rates will rise to decrease aggregate demand and thereby fight inflation, ease the downward pressure on the exchange rate, and allow the real interest rate to regain its normal level.

7. a) Robust job creation suggests that the economy is experiencing a high level of aggregate demand which may create inflationary forces.

b) If the fight against inflation is successful, the value of the U.S. dollar, by PPP, should rise. Further, inflation is usually fought by raising the interest rate. This would induce capital inflows, raising the value of the U.S. dollar. Speculators would buy the U.S. dollar to profit by the capital gain the rise in the value of the dollar would create.

9. . . . the weaker dollar does not lead to cost-push forces that are validated by the monetary authorities, increasing prices.

G. Interest Rate Parity

1. a) The market must be expecting that during the lifetime of these securities the value of the mark will appreciate against the dollar by about 3 per cent per year.

b) The U.S. Treasury must think the mark will appreciate by less than 3 per cent, whereas German investors must think it will appreciate by more than 3 per cent (otherwise they would borrow and lend in dollars).

3. a) . . . higher interest rates.

b) The interest rate in a small open economy such as Canada's is determined by the world interest rate. Excessive borrowing on our domestic market will entice capital inflows as soon as our interest rate rises above the world rate, preventing any substantive rise in our real rate of interest.

5. This is true only of real rates. The nominal rate could remain permanently below that of the U.S. if inflation in Canada were lower than inflation in the U.S. and our exchange rate were allowed to continually rise.

7. The statement is true insofar as it refers to the real interest rate, but if Canada is successful in lowering her inflation rate below that of the U.S., and allows her exchange rate to flex upwards, it is possible to have a fall in the nominal interest rate.

9. If interest rates in Canada are 12 per cent and in the U.S. are 10 per cent, it appears that an investor can gain an extra 2 per cent by investing in Canada rather than in the U.S. But this does not account for the possibility that while one's funds are in Canada, the Canada/U.S. exchange rate might change unfavorably, eliminating part or all of this extra return. Insurance to "hedge" against this risk costs about 2 per cent at present ("the discount on the forward C$ has kept step with the interest rate differential"), so the only way an investor can capture the full difference between the two interest rates is not to buy this insurance (i.e., invest on an unhedged basis).

H. Interest Rates and the Exchange Rate

1. a) When the dollar is strong, a foreigner must pay more of his or her currency to obtain a Canadian dollar, making our exports expensive to them.

b) High interest rates would cause capital inflows which would bid up the value of the Canadian dollar.

3. a) Capital inflows would alleviate pressures on interest rates by augmenting the supply side of the market for loanable funds.

b) Deficits cause the demand for loanable funds to rise. Restrictive monetary policy restricts the supply side of the market for loanable funds. Anticipated inflation directly increases the nominal interest rate.

5. . . . high interest rates. The Bank of Canada has decided to try to lower the Canadian interest rate, knowing that this will involve a fall in the Canadian dollar.

7. a) It would be inflationary if it involved an increase in the rate of growth of the money supply, which is probably what Wilson would have had in mind by lowering interest rates.
   b) A higher inflation in Canada should over time lead to a fall in the value of the Canadian dollar. People would sell off Canadian dollars to avoid suffering this loss.

9. a) If our interest rates fall, foreigners will move their funds to other countries, selling Canadian dollars on the foreign exchange market to do so. This pushes the value of the Canadian dollar down.
   b) Increasing the supply of money should cause its price to fall.
   c) A rise in the money supply suggests that inflation will soon follow.
   d) Unsuccessful. The real rate of interest is set on world markets; the action in question will only serve to activate inflationary forces and will in the long run only raise the nominal interest rate.

11. b) High interest rates, by themselves, increase the demand for the Canadian dollar and push its value up, not down.

I. Policy

1. In an open economy, at each stage of the multiplier process a larger fraction of extra spending is on imports, reducing the multiplier.

3. An upward movement in the Canadian interest rate.

5. If we have zero inflation, but our trading partners do not, with a fixed exchange rate prices of foreign goods will rise relative to prices of our competing goods, so our exports will rise and our imports will fall, creating a balance of payments surplus. This will put upward pressure on our exchange rate, just as PPP predicts. Our exchange rate will have to be allowed to rise at a rate equal to the difference between our trading partner's inflation rate and our (zero) inflation rate.

7. a) Slowing domestic demand would slow import demand, reducing the trade deficit.
   b) Eliminating the trade deficit by allowing the U.S. dollar to fall.

9. If the exchange rate is flexible, borrowing abroad would not increase the money supply, so the stimulus to the economy would not be as great, implying less inflationary pressure. The exchange rate would flex upward, thereby decreasing aggregate demand. If the exchange rate is fixed, both methods will give rise to the same change in the money supply and inflationary pressures.

J. Forecasting the Exchange Rate

1. People were buying Canadian dollars (to reap the benefit of the higher interest rate in Canada) and at the same time selling Canadian dollars on the forward market (i.e., for future delivery, when the bonds they bought mature). The difference between the two countries' interest rates was greater than the difference between the spot and forward rates, so these people would be assured of a profit on this "arbitraging" activity.

3. Businesspeople who have anticipated receipts or expenses in foreign currencies should insure themselves against the possibility of unfavorable changes in the exchange rate.

K. The J Curve

1. The J curve.

II-9A The Natural Rate of Unemployment

1. Draw the long-run Phillips curve as a vertical line at 8 per cent unemployment. Any movement below 8 per cent unemployment will be accomplished by an unexpected increase in the inflation rate, moving the economy up to the left on a short-run Phillips curve. Thus the modern interpretation of the Phillips curve has the higher inflation causing the higher unemployment, rather than the reverse as is implied in this article.

2. The Bank is pursuing a restrictive monetary policy to combat the inflation it believes will accompany the decline of the unemployment rate below its natural rate.

3. Hayes must not believe that Canada's natural rate of unemployment is greater than that of other countries, or greater than what it used to be in Canada.

4. One change in the labor force could be a movement toward more participation by females and young people who have a different level of skills than older males. An example of a change in labor laws could be legislation affecting the strength of labor unions or a change in the generosity of unemployment insurance compensation payments.

5. Non-accelerating-inflation rate of unemployment. To maintain a rate of inflation less than the NAIRU, policy-makers must continually accelerate the rate of inflation to fool people continually.

6. A more generous unemployment compensation system in Canada may be part of the reason. Another may be that more of Canada's labor force is unionized.

7. The participation rate of females may have stopped rising. The cohort of young people joining the labor force may be smaller than in the past.

II-9 Short Clippings

A. The Natural Rate of Unemployment

1. The natural rate of unemployment, or the non-accelerating-inflation rate of unemployment.

3. a) The natural rate of unemployment, or the non-accelerating-inflation rate of unemployment.
   b) Acceleration of inflation.

5. a) Keynes.
   b) On a Phillips curve diagram, it is not possible, in the long run, to move the economy to a rate of unemployment below the natural rate.
   c) By shifting the long-run (vertical at the natural rate of unemployment) Phillips curve to the left.

B. Interpreting the Phillips Curve

1. Any movement to a rate of unemployment below the natural rate requires that workers be fooled into thinking that their current real wage rate is higher than it actually is. This can be accomplished by changing the inflation rate from its current, expected level.

3. Hoarding of labor could be relevant here. For example, suppose that when entering a downturn firms are reluctant to lay off employees in the short run because they do not wish to run the risk of losing employees trained and experienced in their business. With the same number of employees but producing less

output, productivity falls, and there is pressure to raise prices, generating inflation. In general, productivity changes can affect pricing decisions, and thus inflation, without any necessary connection to unemployment in the short run.

5. a) The Phillips curve diagram.
   b) It kills people's expectations of inflation, shifting the short-run Phillips curve down, moving the economy to a lower rate of inflation.

7. a) There was a fixed, downward-sloping Phillips curve. This implied that a movement to a lower (higher) level of inflation was paired with a movement to a higher (lower) level of unemployment.
   b) The link disappeared because people began to expect inflation and build that expectation into their actions.
   c) The economy could have been fighting high inflation with high unemployment. When expectations of inflation finally fall, the short-run Phillips curve will fall, allowing the economy to move to a lower rate of inflation. Then the restrictive policies that were fighting inflation can be relaxed, allowing the economy to move down to its natural rate of unemployment. Alternatively, policies reducing the natural rate of unemployment may have been successful at the same time as policies fighting inflation were successful.

9. a) By demanding a rise in wages to prevent a fall in their standard of living.
   b) Real wages are too high.

C. Using the Phillips Curve

1. That fighting inflation causes a lot of unemployment before that unemployment has much impact on inflation.

3. That a prolonged recession is needed to kill inflationary expectations.

5. Early use of easy monetary policy moved the economy up to the left on the short-run Phillips curve and thus appeared successful in its attempt to achieve a lower unemployment rate. After people adjusted their expectations and the economy moved back to its natural rate of unemployment, the policy authorities, encouraged by their initial success, applied another stimulating dose of monetary policy. This process was repeated, each time accelerating the rate of inflation.

7. Suppose the government is committed to maintain the level of employment at the natural rate. Ordinarily, business and labor would hesitate to increase prices and wages by more than the current rate of inflation because sales would fall and unemployment would develop. But if the government is targeting on this unemployment rate, as soon as unemployment rises above the natural rate it will stimulate the economy, removing this natural constraint on price and wage behavior. This will cause the economy to move quickly up the long-run Phillips curve; the policy invites inflation.

D. Policy for Stagflation

1. By restrictive policies the economy is moved out along the short-run Phillips curve to the right of the natural rate of unemployment (the long-run Phillips curve). Eventually this high level of unemployment will kill expectations of inflation, shifting the short-run Phillips curve down. The economy can then move back to the long-run Phillips curve but now at a lower rate of inflation.

3. If a policy to reduce inflation lacks credibility, expectations of inflation will not fall and the short-run Phillips curve will not shift down. This keeps the economy at a high level of unemployment (the side effect) until the policy becomes credible.

5. If a central bank has a reputation for accomodation, when faced with an infla-

tionary shock business and labor can raise prices and wages without fear of sales reductions and unemployment; this will jump the economy to a higher rate of inflation. On a Phillips curve the inflationary shock will move the economy from the natural rate of unemployment up to the left on the short-run Phillips curve. With an accomodating policy, expectations of inflation will rise and the short-run Phillips curve will shift up; the economy will move eventually to a higher rate of inflation at the natural rate of unemployment. With a non-accomodating policy inflation expectations will not change, the short-run Phillips curve will not move, and the economy will eventually return to its original position.

7. Fixing the nominal interest rate, fixing the exchange rate, or fixing the rate of unemployment at a level below the natural rate.

E. Wage-Price Controls/Guidelines

1. a) Prices and wages are rising in spite of excess supply in their respective markets.
   b) One reason that prices and wages are rising in the face of excess supply is that everyone expects them to rise and so each individual decision to raise a price or a wage is undertaken on the assumption that it can be done without harm because everyone is doing it.

3. a) Inflation would eventually result.
   b) The monetarists feel that decreasing the rate at which the money supply is growing is a sufficient policy solution for inflation. This group feels that this policy will not operate quickly enough and needs to be supplemented with additional policy such as controls.

5. a) The prices Canadian firms pay for foreign inputs cannot be controlled, so it will be difficult if not impossible for them to keep their costs below the controlled price.
   b) These uneven changes imply that the economy is trying to readjust resources through relative price changes. This is the mechanism which keeps the economy producing the right goods in the right quantities in the most efficient fashion. If controls were applied these desirable changes in relative prices would be prevented.

F. Supply-Side Economics

1. Fiscal policy and supply-side policy.

3. a) From the demand side it increases after-tax income, increasing spending. From the supply side it increases the after-tax return to work and investment, stimulating both.
   b) Laffer's statement explains how a fall in the tax rate could bring about a net increase in tax receipts. At a low tax rate, however, the incentive effects of the tax would be weak and thus a fall in the tax rate would lead to a net fall in tax receipts. This implies that a hump-shaped figure results when tax receipts are graphed against the tax rate.

5. Shift the supply curve to the right.

7. a) Business will not invest in new capital stock (capacity) because it is not needed; the capital stock may even shrink due to depreciation.
   b) When aggregate demand recovers, capacity constraints will prevent producers from increasing aggregate supply to meet it, resulting in inflation at a high level of unemployment.

9. Increase the money supply at a low, fixed rate and wait through the resulting unemployment until inflation subsides.

II-10B Costs vs Benefits of Deficit

1. When the government sells bonds to finance its deficit, it bids up the interest rate to obtain those funds, causing some private investment (in plant and equipment) to be forgone as it is no longer profitable. By crowding out this private investment, the income that this investment would have generated is lost.

2. In the case described in this paragraph, in which people respond to an increase in the deficit by saving more, investment financed by Canadians will be unaffected regardless of whether the deficit is financed by raising taxes or by selling bonds. In each case the government financing comes entirely from higher saving, implying no crowding out of the existing level of investment. As well, in each case the fall in consumption is identical.

3. Inflation. Those "paying" would be those unable to protect themselves from the ravages of inflation.

4. Government borrowing will bring in capital inflows instead of causing a marked increase in the interest rate. This will bid up the value of the Canadian dollar causing the crowding out to take the form of a decrease in exports and a decrease in import-competing goods and services.

II-10 Short Clippings

A. Monetizing the Debt

1. The money supply will be increased if the government borrows from the Bank of Canada.

3. The deficit could be financed by printing money (instead of by raising taxes), which would accelerate inflation.

5. The Bank could refuse to print money to help finance the government's deficit. If it does not decrease its deficit, the government would have to sell more bonds, resulting in a higher interest rate.

B. Deficits and Interest Rates

1. a) Funds that could be used in the private sector must be used to finance the deficit.
   b) The interest rate rises.
   c) By printing money. This runs the risk of raising inflation.

3. It probably would not be able to sell its bonds at the lower interest rate.

5. The high interest rates may have been caused by the high deficit. The deficit spending may have created inflation, especially if some of it was financed by printing money, and this inflation would raise the nominal interest rate.

C. Crowding Out

1. Through a fall in demand because of the higher interest rate occasioned by the deficit; through a fall in consumption (rise in saving) by those worried about higher future taxes because of the deficit; and through a fall in demand for exports and import-competing goods due to the rise in the exchange rate brought about by higher foreign borrowing.

3. If the federal government did not need so much financing, others could obtain their financing at a lower interest rate. The interest rate is what determines whether Ottawa or others get the cash—Ottawa just bids up the rate so that others are squeezed out, allowing Ottawa to get the cash it wants.

5. Reducing federal deficits should imply that more financing is available for the private sector to undertake productive investments. But if inflation persists, the

private sector may choose to use this financing for unproductive purposes designed to protect itself against inflation.

7. a) A reduction of the deficit shifts the demand curve for loanable funds to the left, reducing the interest rate.
   b) The answer only makes sense if private demand for loanable funds is infinitely elastic at the current interest rate, an unlikely assumption.

D. Deficits and the International Dimension

1. To finance the deficit the government must sell bonds at an attractive interest rate. This entices foreigners to invest in our bonds, and to do so they buy Canadian dollars on the foreign exchange market, pushing up the price of Canadian dollars. This causes a fall in Canadian exports.

3. By arguing that the interest rate is determined internationally, and that capital will flow into Canada to prevent the interest rate from rising too much above the world rate.

5. The U.S. is so large that its activities will have a noticeable impact on interest rates. If other countries curb their deficits, their governments will not be selling so many bonds and so in their bond markets the interest rate should fall. As soon as it attempts to do so, however, capital will flow out of that country in search of the higher rate of return available elsewhere (mainly in the U.S.). This outflow reduces the funds on the supply side of this bond market and thus causes the interest rate to move back up in line with the world interest rate.

E. Structural Deficit

1. The greater the economy's growth, the greater the tax revenues, and consequently the smaller the deficit.

3. . . . the causation is the reverse of what this argument claims. It is the high unemployment which is reducing tax receipts and increasing transfer payments, that has caused the high deficit, rather than the other way around.

5. a) Compute what tax receipts would be if the economy were fully employed, then take the difference between this figure and current government spending to get the full employment deficit. The current deficit would be larger.
   b) Deficits were only supposed to be used to move the economy out of recessions; they were to be balanced off by surpluses during inflationary times.

F. Keynesian Policy

1. Tax hikes or expenditure cuts would decrease aggregate demand and thereby move us towards recession.

3. It suggests that deficit government spending is an appropriate policy in recession.

5. Spending should be financed by taxes, so a decision to spend can be a decision to tax. But spending can be financed by selling bonds, which will increase the national debt, or by printing money, which is inflationary, so a decision to spend is only a decision to tax if we wish to avoid these two alternatives.

II-11 Short Clippings

1. The rise in the Canadian dollar makes Canada less competitive relative to the U.S.

3. If wage costs in Canada become higher than in the U.S. and as a result Canada's exports fall, the resulting fall in the exchange rate would restore export

sales. In effect, the change in the exchange rate causes Canada's costs, in U.S. dollar terms, to return to their original level, maintaining a competitive position.

5. If the Canadian dollar is depreciated, the cost of imports and anything with import content will rise, decreasing the purchasing power of our wages. An advocate of depreciation would respond to this by saying that the current level of real income in Canada is too high for us to sustain full employment; the fall in real income is the cost of getting the economy to move to full employment.

Glossary

Ad Valorem Tax—a tax on a good or service which is a fixed percentage of the value of the good or service, e.g., a sales tax.

Arc Elasticity—elasticity calculated using averages of the before and after figures as bases for calculation of percentage changes.

Automatic Stabilizer—nondiscretionary or "built-in" features of government revenues and expenditures that automatically cushion recession by helping to create a budget deficit and curb inflation by helping to create a budget surplus, e.g., unemployment insurance payments and benefits, income tax, and agricultural price supports.

Average Propensity to Consume (APC)—the ratio of consumption to income.

Average Total Cost (ATC)—total cost divided by the number of units of output.

Average Variable Cost (AVC)—total variable cost divided by the number of units of output.

Balance of Payments—the difference between a country's annual supply of foreign exchange and its annual demand for foreign exchange.

Balance of Trade—the difference between a country's exports and imports (of goods).

Balanced-Budget Multiplier—the multiplier associated with an increase (or decrease) in government spending financed by an equal increase (decrease) in taxes.

Bank of Canada—Canada's central bank.

Bank Rate—the interest rate at which the Bank of Canada will loan cash reserves to the chartered banks. The comparable U.S. rate is called the *discount rate.*

Bellwether—a signalling device.

Bottleneck Inflation—*see* Demand Shift Inflation.

Bretton Woods—site of a 1944 international monetary conference at which the postwar world fixed exchange rate system was structured and the International Monetary Fund (IMF) was created.

Broad Definition of Money—*see* Money.

Capital—a factor of production, defined to include all man-made aids to production such as buildings, equipment, and inventories but not stocks, bonds, mortgages or money.

Capital Account—the difference between a country's annual capital inflows and capital outflows.

Capital Inflows—a supply of foreign exchange to a country arising from foreigners purchasing that country's financial assets or businesses. This would be a *capital outflow* to the foreign countries in question.

Capital Stock—the aggregate quantity of an economy's capital goods.

Capitalize—calculate a lump sum equivalent to a series of payments.

Cartel—an organization of producers designed to limit or eliminate competition among its members; the term generally applies to government-enforced agreements.

Cash Reserves—see Reserves.

Central Bank—a bank that acts as banker to the banking system and to the government. It is the sole money-issuing authority.

Ceteris Paribus—literally, "other things being equal." Used to indicate that all variables except those specified are assumed unchanged.

Chartered Bank—a bank licensed as such by Parliament under the Bank Act; they are subject to considerable regulation by the Bank of Canada. There are only about a dozen chartered banks in Canada. The comparable U.S. term is *commercial bank* (of which there are thousands).

Cobweb—*See* Corn-Hog Cycle.

Commercial Paper—short-term debt issued by well-known corporations.

Complement—goods used in conjunction with one another are complementary goods (ham and eggs); if the price of one good rises (falls) the demand for the other good falls (rises).

Consumers' Surplus—the difference between the amount paid for a commodity and the highest amount the consumer would be willing to pay to avoid going without.

Corn-Hog Cycle—a repetition of the sequence of low output and high prices followed by high output and low prices. These cyclical fluctuations in price and quantities arise for certain agricultural products where the quantity supplied to the market is determined at a previous time when production plans have to be formulated.

Corporatism—a system in which labor, business and government leaders meet regularly and negotiate the division of national income.

Cost-Benefit Analysis—an attempt to estimate the total costs and total benefits to society of public programs or policies.

Cost-Push—a theory of inflation that rests on cost increases caused by factor payments increasing faster than productivity or efficiency: This inflation is unrelated to excess aggregate demand.

Crowding-Out—decreases in private aggregate demand associated with the means used to finance an expansionary fiscal policy.

CSB—Canada Savings Bond.

Current Account—the balance of trade augmented by the net foreign exchange transactions associated with services.

Demand Deposit—a bank deposit that is withdrawable on demand and transferable by means of a cheque.

Demand-Pull—a theory of inflation that rests on excess aggregate demand.

Demand-Shift Inflation—a theory of inflation that rests on a lack of matching of the components of aggregate demand and aggregate supply.

Depreciation—the decline in the value of capital due to its wearing out or becoming obsolete.

Depression—a prolonged period of very low economic activity which is generally characterized by large-scale unemployment of resources.

Devaluation—a drop in the value of a country's currency on international exchange markets (or a rise in the cost of buying foreign exchange).

Dirty Floater—a country whose government buys or sells foreign exchange so as to influence the market-determined value of its currency.

Dumping—sale of the same product in different markets at prices lower than in the "home" market; generally associated with selling a good in a foreign country at a price below the domestic price.

Dynamic—relating to movement over time. Rather than simply comparing old and new equilibrium positions (comparative statics), a dynamic analysis examines the character of the economy's movement from the old to the new position.

Economies of Scale—a decrease in a firm's long-run average costs as the size of its plant is increased.

Efficiency—the degree to which the economy achieves a position in which it is impossible to reallocate or redistribute goods and services so as to make someone better off without making someone else worse off.

Elasticity—*see* Price Elasticity.

Envelope—a curve formed by joining the extremities of a family of curves.

Equilibrium—a state of balance in which there are no internal pressures for change, usually characterized by equality between demand and supply.

Excess Demand—demand exceeding supply at a given price. The reverse is called excess supply.

Exchange Rate—the price of one currency in terms of another.

Exogenous—autonomous, not originating from within the system.

External Diseconomies—by-products (costs) of a firm's production process for which it does not pay.

External Economies—by-products (benefits) of a firm's production process for which the firm is unable to charge.

Federal Reserve Board—the central bank of the United States.

Fiscal Policy—a change in government spending or taxing, designed to influence economic activity.

Fixed Costs—costs that do not change as the firm changes its level of output.

Fixed Exchange Rate—an exchange rate held at a constant level by a government promise to buy or sell foreign exchange at that rate.

Flexible Exchange Rate—an exchange rate whose level is determined by the unhindered market forces of supply and demand for foreign exchange.

Floating Exchange Rate—*see* Flexible Exchange Rate.

Foreign Exchange Fund—a fund containing the government's holdings of foreign currency or claims thereon.

Forward Exchange Market—a market in which foreign exchange can be bought or sold for delivery (and payment) at some specified future date but at a price agreed upon in the present.

Frictional Unemployment—unemployment associated with people changing jobs or quitting to search for new jobs.

Futures Market—a market in which a good can be bought or sold for delivery (and payment) at some specified future date but at a price agreed upon in the present.

GNP—gross national product, the total of all final goods and services produced during the year.

Hedging—a technique of buying and selling future contracts that minimizes the risk of loss due to price fluctuations.

Hog Cycle—*see* Corn-Hog Cycle.

Indifference Curve—a graph of combinations of quantities of two goods, with any of which a consumer is equally satisfied; each of these combinations yields the same total utility to the recipient.

Interest Rate Parity—in the absence of restrictions on capital flows, real interest rates will tend to be the same in all countries.

Labor Force—the number of people either employed or actively seeking work.

Liquidity—the ability to meet current financial liabilities in cash.

Liquidity Preference—a theory of the determination of the interest rate, resting on the supply of and demand for money.

Loanable Funds—a theory of the determination of the interest rate, resting on the supply of and demand for bonds and other financial assets.

M1, M2—*see* Money.

Marginal Benefit—the extra benefit created by increasing or decreasing output (consumption) by one unit.

Marginal Cost—the extra cost resulting from increasing output by one unit.

Marginal Propensity to Consume—the proportion of an extra dollar of income that is spent on consumption of goods and services.

Marginal Revenue—the change in a firm's revenue resulting from selling one additional unit of output.

Marginal Utility—the additional satisfaction gained (or lost) by a buyer from consuming one unit more of a good.

Marketing Board—an association of producers set up, under government sanction, to market output jointly.

Mixed Economy—an economy in which some decisions as to what to produce, and for whom to produce, are made by firms and households and some by central authorities.

Monetarists—economists who believe that fluctuations in the quantity of money are the primary causes of economic fluctuations.

Monetary Policy—a change in a monetary variable, such as the quantity of money, designed to affect economic activity.

Money—the narrow definition of the money supply (M1) is the sum of currency and demand deposits. The broad definition (M2) is M1 plus time deposits. *See also* Printing Money.

Money Market—the market in which money is borrowed and loaned; the stock and bond market.

Money Rate of Interest—the observed interest rate in the money market.

Monopolistic Competition—the structure of an industry with many sellers and freedom of entry, each seller having a product slightly different from the other, giving him some control over his price.

Monopoly—an industry characterized by a single seller.

MPC—see Marginal Propensity to Consume.

Multiplier—the amount by which national income increases when government spending increases by one dollar. In more general terms, the factor by which the magnitude of a policy action must be multiplied to give the impact of that policy on some specified dimension of economic activity.

Narrow Definition of Money—*see* Money.

Natural Rate of Unemployment—the unemployment rate to which it is thought the economy will gravitate over the long run. Its magnitude is determined by institutional factors, such as the degree to which changes in tastes and technology maintain structural unemployment, the degree to which information problems and geographical immobility maintain frictional unemployment, and the levels of the minimum wage and UIC benefits.

Nominal Rate of Interest—see Money Rate of Interest.

Oligopoly—a market structure in which a smaller number of rival firms dominate the industry.

OPEC—Organization of Petroleum Exporting Countries.

Open Economy—an economy that engages in foreign trade.

Open-Market Operations—the purchase or sale of securities on the money market by the central bank with the goal of expanding or contracting the supply of money and credit.

Opportunity Cost—the return a resource could earn in its most profitable alternative use.

Participation Rate—the percentage of the noninstitutionalized population over age 15 that is in the labor force.

Peripheral Work Force—workers who are not the primary breadwinners in their household.

Phillips Curve—a relation between the rate of change of money wages and unemployment, often drawn as a relation between inflation and unemployment.

Price Discrimination—the sale by a firm of the same commodity to different buyers at different prices, for reasons unrelated to costs.

Price Elasticity of Demand—the percentage change in quantity demanded caused by a price change (*ceteris paribus*), divided by the percentage change in price, conventionally expressed as a positive number.

Price System—the use of freely-determined market prices to direct the allocation and distribution of goods and services within an economy.

Prime Rate—the interest rate charged by chartered banks to their most favored customers.

Printing Money—the sale of bonds by the government to the central bank.

Producers' Surplus—revenue received by producers, in excess of their actual production cost.

Progressive Tax—a tax which, as a percentage of income, falls more heavily on those with higher incomes.

Purchasing Power Parity—in the absence of trade restrictions, traded goods will tend to cost the same whether purchased locally with local currency or imported paying foreign currency.

Quota System—a production system in which each producer is permitted to produce only a specified quantity, called his quota.

Real Income—money income corrected for changes in the price level (the purchasing power of money income).

Real Rate of Interest—the money rate of interest less the expected rate of inflation (the money rate of interest corrected for the expected change in the purchasing power of money).

Real Wage—the money wage corrected for price level changes.

Recession—a downswing in the level of economic activity.

Reflation—the process of managing money for the purpose of restoring a previous price level.

Rent—the return to a factor in excess of the return required to entice that factor to its present employment.

Reserves—in banking, that part of customers' deposits kept by a bank either as currency held in its own vault or as deposits with the central bank.

Sales Tax—a tax levied as a percentage of retail sales.

SDR—special drawing right, the name given to the "currency" of the International Monetary Fund.

Spot—for immediate payment and delivery, as opposed to future payment and delivery.

Stagflation—the simultaneous existence of high inflation and high unemployment.

Sterilization—central bank action offsetting money supply changes brought about, under a fixed exchange rate system, by balance of payments deficits or surpluses.

Structural Unemployment—unemployment resulting from technological displacement, lack of appropriate skills, and geographic imbalances between the supply of and demand for labor.

Substitute—two goods are said to be substitutes if an increase (decrease) in the price of one leads to a rise (fall) in the demand for the other.

Swap Agreements—holdings of U.S. Treasury securities by the Bank of Canada, purchased by the Bank from the government's Exchange Fund Account, with a promise by the government to buy these securities back at a set price.

Tariff—a tax applied on imports which can be based on a tax per unit of the commodity or on the value of the commodity.

Term Deposit—a savings deposit where the depositor specifies the length of time the money is to be deposited. It generally can be withdrawn before its maturity date with a penalty such as a lower interest rate paid. Many writers use term, time and notice deposits interchangeably.

Time Deposit—an interest-earning bank deposit, subject to notice before withdrawal; also called a notice deposit.

Treasury Bill—the characteristic form of short-term government debt.

Tripartism—*see* Corporatism.

UIC—Unemployment Insurance Commission.

Utility—consumer satisfaction.

Variable Costs—costs whose total varies directly with the level of output.

Velocity—the ratio of GNP to the money supply.